BATTLESHIP
AT WAR

Below:
***Princess Royal*, with 'A' turret traversed to starboard, played a major part in the Battle of the Dogger Bank.** *Real Photos 307*

BATTLESHIP AT WAR

Cdr B.R. Coward RN

LONDON
IAN ALLAN LTD

First published 1987

ISBN 0 7110 1740 9

Published by Ian Allan Ltd, Shepperton, Surrey; and printed by Ian Allan Printing Ltd at their works at Coombelands in Runnymede, England

Below:
***New Jersey* (BB-62) engaging targets ashore during operations off the coast of Vietnam.**
US Navy

Contents

Preface

In common with other books in the 'at War' series, the aim of *Battleship at War* is to describe 'what it was like to be there'. Thus the emphasis is on the people rather than the ships. The book does not try to be a definitive account of the major actions in which battleships participated — such accounts are to be found in many other excellent books. However, it seemed logical to structure the book in a chronological sequence and at least to provide an outline narrative of the major events in which battleships participated. Indeed some of the key actions, such as the Battle of the North Cape, have been dealt with in some detail, whilst others, notably the Battle of Jutland, have been examined only very briefly.

As much as is possible I have used the accounts of men who were actually there at the time. In some cases, such men played only a small part, but such was the nature of life in a battleship. The running of the ship and success in battle depended on each member of the large crew playing his small part. Only certain key personnel, such as the Captain, Gunnery Officer, Commander (E) had major roles to play; a small number of ratings in vital positions, such as the main armament director layer and trainer, too could decide the outcome of a particular event. But, on the whole, it was team work that counted, and, whenever possible, I have used the accounts of those team members — the ordinary members of the ship's company.

Having said that, it is a fact that many of the surviving accounts of World War 1 were written by officers. Indeed, the vast majority of them were written by Midshipmen, mainly because Midshipmen were required to write journals, and many of these have survived to this day. I have been able to include a few accounts written by ratings as well. It will be appreciated that very few men who served in battleships in World War 1 are still alive and thus we must work from the accounts that they left behind.

For World War 2, most of the material in this book is new and has been provided by

Right:
Men going to Action Stations on board *Anson*. On the right is the door to the aircraft hangar with the nose of a Walrus amphibian just visible. *IWM A16503*

private individuals. Their accounts cover a wide range, from signalmen to engineers, from gunnery ratings to stores accountants and thus it has been possible to describe a broad spectrum of life on board in the words of those who actually did the job 40-odd years ago.

The nature of the battleship war during World War 1 bore very little resemblance to the theory which had been taught before 1914. The concept of vast fleet actions, of ships slugging it out in gunnery duels, was only put to the test once, at the Battle of Jutland. Certainly, few had even conceived of what was to prove to be the longest sustained operation by battleships, the Dardanelles campaign, where the primary task of the ships was to engage targets ashore. Nevertheless life in battleships, whether in action or not, was hard physical work. Tasks such as coaling ship, ammunitioning and provisioning were very labour-intensive. Despite having power loading on the larger guns, much of the work of operating and firing guns too required sheer muscle power. For the stokers shovelling coal into the hungry boilers, their whole existence was hard physical toil. And as well as being hard work, much of the life was boring. At sea, men spent long hours closed up at their stations, just waiting. In harbour there were few amenities, in particular in places like Scapa Flow, and there was little leave. Life on board, by today's standards, was basic and uncomfortable.

At the start of World War 2 there was a more realistic appreciation of what the war at sea would be like. There would be few, if any, fleet actions. Much of the battleship war would be centred on the protection of the convoys and the bombardment of targets ashore in support of land operations. However, in 1939 few probably appreciated that the battleship war would draw to a close off the shores of Japan.

In 1914, the major threats to battleships came from other battleships and torpedo-firing destroyers. The danger from submarines was underestimated, although it would be the submarine threat which would eventually put an end to battleship operations in the Dardanelles. In 1939 the U-boat threat was well understood but the power of the aircraft was badly misjudged. Few could imagine that two British battleships would be sunk by aircraft alone and that, by the end of the war, the aircraft carrier would have assumed the role as the principal striking arm of the Navy.

This book concentrates mainly on the British battleships at war as that is from where most of the accounts have been drawn. However, some German and French operations are included in the section on World War 1. German operations are covered in greater detail for World War 2, and US operations are described at length in the sections on the Far East and post-1945. Japanese, French and Italian battleships are included for World War 2 also.

Top:
***Warspite* bombarding Brest in the late summer of 1944.** *E. R. Moyle*

Right:
Loading stores in harbour — HMS *Revenge*. *E. R. Moyle*

Above:
One of the 'Queen Elizabeth' class underway in the North Sea during World War 1.
P. Liddle/Vice-Adm E. Langley Cook

Acknowledgements
The author is most grateful to the many individuals who took the time and trouble to record their memories and who provided photographs and other material so readily. The individual contributors are listed: without their help, this book could not have been written. The author is particularly grateful to Mr Peter Liddle of Sunderland Polytechnic who gave access to his considerable archives of 1914-18 material, from which most of the World War 1 accounts have been drawn. Again, without Peter Liddle's help, few accounts of that period would have been available.

Thanks are due to the US Naval Institute Press for permission to quote from *The Japanese Navy in World War II* and to the Imperial War Museum, Maritime Photo Library, US Navy, P. Liddle and the Fleet Air Arm Museum for the provision of photographs. Robert A. Carlisle of the US Navy, Office of Information, was of particular assistance in providing advice on US Naval photographs and in researching material. Photographs provided by private individuals are acknowledged with the captions. The author is most grateful to those who were willing to send him their photograph collections.

Photograph Captions

IWM	Imperial War Museum, London
MPL	Maritime Photo Library, Cromer
P. Liddle	P. Liddle 1914-18 Archives, Sunderland Polytechnic

Author's Note
Wherever possible, I have quoted extracts from individual's accounts verbatim. However, occasionally, a more flowing narrative resulted from editing out short sections. Where explanation is appropriate, I have included it in square brackets within the quotation. Some anecdotes of battleship life in World War 2 do not fit easily into the chronological narrative but deserved inclusion nevertheless. They appear from time to time indicated in italics.

For those who may be unfamiliar with some naval terminology and parlance, I have included a short glossary, and have quoted the names of ships in italics without the prefixes HMS, USS, etc. Also I have quoted times in the convention used in the original material: thus the accounts of World War 1 contain times such as 7am rather than the modern military convention of 07.00.

Glossary

Artificer	A skilled technician
'A' tube	The inner tube of a gun barrel
Block and tackle	Pulley system
Bosun's mate	Junior rating who assisted the Quartermaster in keeping watch at the gangway in harbour
Buzz	Rumour or message
Casemate	Compartment, usually along the ship's side, in which a number of guns of the secondary armament were mounted
CRUDIV	Division of cruisers
DESDIV	Division of destroyers
Dhobeying	Washing
Fanny	A container for liquid, usually of several gallons capacity
Flat	An open area below decks
Goffer	Soft drink
Guide	The ship from which other ships take their position in a formation
Handsomely	Slowly, with care
HA	High-angle
Howitzer	A gun designed to fire at a high angle of elevation so that its shells plunge on to the target
Iron deck	Upper deck
Kye	Hot chocolate
LA	Low angle
Lighter	Barge
Master at Arms	Senior rating responsible for discipline
Nutty	Sweets, bars of chocolate, etc
Oerlikon	20mm anti-aircraft gun
Ordnance Artificer (OA)	Technician responsible for the maintenance of the guns
Plumber	Engineer
Point	22½°, from points of the compass
Pom-pom	Anti-aircraft gun firing 2pdr shells
Pusser	Supply Officer, but also used to mean anything belonging to the Navy, eg Pusser's rum
QF	Quick firing (guns)
Quartermaster	Rating who stands watch at the gangway in harbour and in the wheelhouse at sea
Scran bag	Bag into which items of clothing left lying around on messdecks are placed, the payment of a small fine being necessary to retrieve them
VA	Vice Admiral
Waist	The amidships area of a ship
Wrinkles	Hints, tricks of the trade

WORLD WAR I

One

World War 1 – The Beginning

1 Early Days

When war broke out in August 1914, the Royal Navy possessed 76 battleships and battlecruisers. Of those, 41 had been built before the revolutionary *Dreadnought*, and were known collectively as 'pre-Dreadnoughts'. The remainder were altogether more modern, incorporating the basic ideas of *Dreadnought*, plus improvements made since *Dreadnought* herself was commissioned in 1906. The most modern class of battleship at the beginning of the war was the 'Iron Duke' class. These ships mounted 10×13.5in main armament and 12×6in secondary armament, had armour up to 12in thick in the main belt and 11in in the turrets, were propelled by steam turbines driving four screws for a maximum speed of about 21kt, and had a complement of some 1,000 men. They were, however, still coal burners.

At the other end of the spectrum were the 'Majestic' class ships built in the 1890s, mounting only 4×12in as the main armament with 12×6in guns in the secondary battery, and limited to speeds of 16-17kt by their triple expansion engines. And in classes of their own were the battlecruisers, with armament similar to the battleships but more lightly armoured and with more powerful main machinery, the latest, *Tiger*, being capable of 29kt.

The majority of these ships were in home waters but four were employed on guardship duties at ports around the British coast and 11 were stationed abroad. The majority of the ships joined the Grand Fleet based in the north of Scotland, although some of the

Below:
Part of the Grand Fleet going to sea: the 1st and 2nd Divisions of battleships leave the Pentland Firth. *MPL*

Bottom:
When completed in 1914, *Tiger* was the heaviest ship to join the Grand Fleet. Her main armament comprised 8×13.5in plus 12×6in guns; she was capable of 29kt. *MPL*

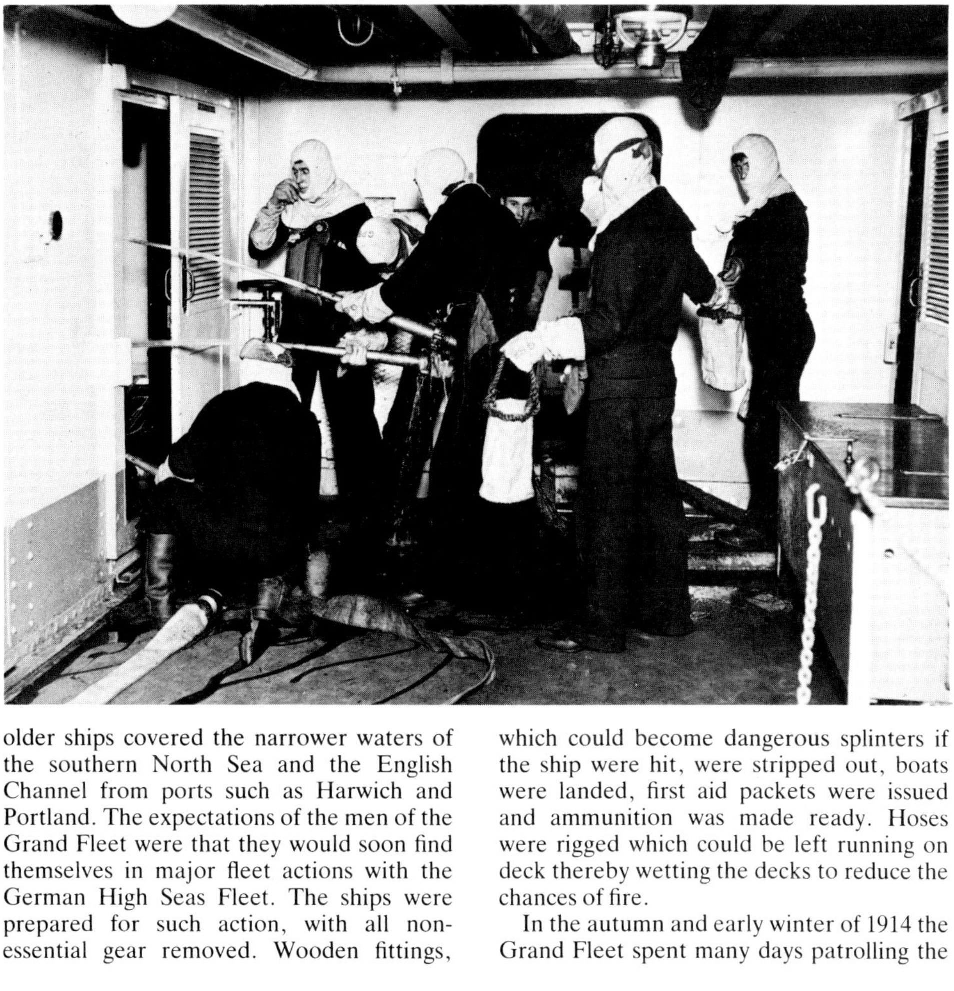

Above:
The 1st Battle Squadron in the North Sea in April 1915. However, the North Sea was not often as placid as in this picture. *IWM SP519*

Left:
Fire drill onboard a battleship; the men are seen wearing anti-flash hoods and gloves.
IWM Q18003

older ships covered the narrower waters of the southern North Sea and the English Channel from ports such as Harwich and Portland. The expectations of the men of the Grand Fleet were that they would soon find themselves in major fleet actions with the German High Seas Fleet. The ships were prepared for such action, with all non-essential gear removed. Wooden fittings, which could become dangerous splinters if the ship were hit, were stripped out, boats were landed, first aid packets were issued and ammunition was made ready. Hoses were rigged which could be left running on deck thereby wetting the decks to reduce the chances of fire.

In the autumn and early winter of 1914 the Grand Fleet spent many days patrolling the grey wastes of the North Sea. The Fleet was seldom at its base at Scapa Flow for more than one or two days to coal and provision before returning to sea. The men started to learn the realities of war — life was 90% boredom, cold, lack of sleep and continual strain: they spent many hours closed up, day and night. The nights were particularly trying. Picture the scene in a gun battery — pitch dark except for a shrouded blue light over the breech of each gun, dim shapes of men muffled against the cold, whispered conversation and the sounds of the eternal wind and sea. Secondary armament was manned to guard against night destroyer attack. At first men were not allowed to sleep at their stations, but this was relaxed later so that some slept whilst others remained alert.

What this early period did allow was that the ships of the Grand Fleet became used to working with each other. Ships practised manoeuvring in close formation, they practised signalling, they practised tactics. They carried out gunnery exercises, torpedo exercises, machinery breakdowns, fire fighting, damage control. They became worked up as a Fleet.

2 The Battle of the Dogger Bank

Although a major fleet action was not in fact to take place until 1916, the RN's battlecruisers were involved in a skirmish with German cruisers as early as 28 August 1914 in what became known as the Battle of the Heligoland Bight. A few months later, in January 1915, the battlecruisers under Adm. Beatty this time met the four German battlecruisers *Seydlitz*, *Derfflinger*, *Moltke* and *Blucher* off the Dogger Bank. J. G. D. Ouvry, then a Midshipman in *Tiger*, describes the day:

'On the morning of the 24th, I turned out at 5.30am. I had my breakfast at 6. At 7 o'clock General Quarters were sounded off. The light was pretty good and the day clear and calm . . . The Captain was up on the bridge as were the Engineer Captain and the Navigating Commander. I was looking out for submarines and mines on the starboard side. We were to come up with three destroyer flotillas at 7.30am at a settled rendezvous. At 7.20 we had met two flotillas which were then on either side of us. Suddenly flashes were seen on our port bow. An engagement appeared to be in progress between some of our light cruisers and the enemy's cruisers. This firing died down a few minutes later. Following this, our speed was increased to 22 knots and later increased again up to 26. From this speed we increased gradually up to 29 knots. At 7.30am we sighted the smoke of 7 hostile ships on our port bow, which were the enemy's battlecruisers and some light cruisers. The Flagship then signalled to the squadron to form quarter line disposed to port. We were then at the rendezvous and all our destroyers were there. Flotillas and light cruisers were sent ahead of us, thus hiding us with their smoke. Many trawlers were steaming about at the time and a sharp lookout had to be

kept on them in case they were concealing submarines, or laying mines. Meanwhile our speed was being increased and soon the destroyers dropped back and the stern chase began. Our speed reached 29 knots. Only the *Lion*, *Tiger* and *Princess Royal* could keep up. *New Zealand* kept fairly well up, but *Indomitable* was forced to drop back amongst the destroyers and light cruisers.

'At 9 o'clock the enemy ships were fairly clearly outlined. They were, however, to windward and thus smoke covered them, while we were exposed. The range had then decreased to roughly 20,000 yards. *Lion* then fired a ranging shot with one of her 13.5in guns. We followed her example. The Captain and other officers on the bridge then transferred their position to the conning tower. I and the other Midshipman were left outside as submarine lookouts, alongside the 3pdrs. The *Lion* now opened fire for good as did we. The shots appeared to be falling short. The enemy also opened fire, their shots falling about 500 yards short. The Captain now ordered the two of us into the conning tower to lookout there.

'Afterwards some of the crew rescued from the *Blucher* admitted that the *Tiger's* first shot went down one of their funnels, projectiles at this range falling almost vertically on to the ship. The enemy concentrated almost all their fire on the *Lion* and the remainder on the *Tiger*. Their opening salvo seemed pretty good and both ships were straddled by their salvoes. Our firing seemed to be fair, our director system, the first ever fitted, being very prominent . . . Meanwhile our firing was improving and the *Blucher*, being badly hit, was forced to drop behind the remainder. *Derfflinger* also was seen to be on fire. A Zeppelin then appeared over the *Blucher* so our A turret was detailed off to keep an eye on it.

'[After *Lion* had dropped out of the action] the *Tiger* now took guide of the fleet

Above:
***Lion* was Adm Beatty's Flagship at the Battle of the Dogger Bank until she was forced to drop out of the action after being damaged by German gunfire.** *IWM SP1316*

Below:
Battlecruisers at anchor in Scapa Flow. The three ships of the 'Invincible' class are nearest the camera, the fourth ship being of the 'Indefatigable' class; all carried 8×12in guns. Three funnels was the recognition feature of the early battlecruisers. *MPL*

and the enemy's fire was now concentrated solely on the *Tiger*. Shells fell thick and fast all round us, though throughout only about 8 shells struck us. 2 or 3 destroyers now came up and were fired at by the *Blucher*. They, however, were not hit. The enemy destroyers made feeble efforts to attack us from the port bow but we opened fire with our 6in and they soon turned and made off. We had a very exciting time in the conning tower. One could now see the enemy's 7 ships plainly . . . Our fire had been shifted from one ship to another, whichever offered the clearest mark, but now we concentrated on the *Derfflinger*. Both she and the *Moltke* were badly on fire. A report came through that the director was out of action so turrets were put into secondary control from B turret. Shortly came a report that a shell had hit the seam of the roof of Q turret — both guns were out of action. One could see the brilliant flashes of the enemy's guns. One wondered if any of them were going to be hits. First you could see a number of shells pitched slightly short, next time a little nearer, then it seemed the next must hit.

'Suddenly there was a loud explosion and all in the conning tower felt a shock. Smoke came pouring in through the lookout slit. However, everyone kept perfectly cool and no-one was hurt. Soon came the news that a shell had pierced the 3in armour of the "Rowton House" and had passed through the deck to the intelligence office where it had exploded. The bakery was set on fire as was the intelligence office. (The Rowton House was the rangefinder hand-training compartment beneath the conning tower and gunnery control tower.)

'Our ships were now approaching the proximity of the enemy's submarines and minefields and so we made an 8 point turn [90°] thus cutting off the *Blucher*. The remaining enemy's battlecruisers managed to escape, *Derfflinger* and *Moltke* being in flames. We now steamed round the *Blucher* which, by this time, had almost stopped. The Zeppelin had drawn off. Several destroyers closed on the *Blucher* and she fired her after guns at them . . .

'By this time she had stopped and was little more than a wreck. She still discharged

Below:
***Blucher*, the only German battlecruiser to be sunk at the Battle of the Dogger Bank.**
P. Liddle/Cdr J. G. D. Ouvry

solitary guns at us. Soon she ceased fire altogether. One of our salvoes exploded by her A turret and the whole turret was blown over the side. Another blew away the conning tower and bridge.'

Lion, Beatty's flagship, was hit 12 times and had to pull out of the action, flooded with 3,000 tons of water. *Tiger* received two hits. On the German side, *Blucher* was sunk, *Seydlitz* badly damaged, *Derfflinger* was hit once with minor damage and *Moltke* escaped unscathed.

In home waters, major warships were not to clash again for another 16 months. However, further afield, other battlecruisers had also been in action.

Left:
The filth and grime of a coal-fired boiler room. The 'Lion' class had 42 boilers; when they were steaming at maximum speed the stokers faced a herculean task keeping the boilers fed with coal. *IWM Q18594*

3 The Battle of the Falkland Islands

In August 1914 the German East Asiatic Squadron, commanded by Vice-Adm Graf von Spee, consisted of the armoured cruisers *Scharnhorst* and *Gneisenau* plus the three light cruisers *Emden*, *Leipzig* and *Nurnberg*. *Emden* was subsequently destroyed by the Australian cruiser *Sydney* off the Cocos Islands on 9 November 1914. Meanwhile, Von Spee, after ineffectual raids in the mid-Pacific, concentrated the other ships of his force off the west coast of Chile in the middle of October, at which time he was joined by the light cruiser *Dresden*. His intention was to attack merchant shipping in the area.

Protecting British trade off Chile was a squadron under the command of Rear-Adm Sir Christopher Cradock, consisting of the 15 year-old battleship *Canopus*, armed with 4×12in guns, the armoured cruisers *Good Hope* and *Monmouth*, the light cruiser *Glasgow* and the armed merchant cruiser *Otranto*. On 1 November Cradock's force, less *Canopus* which was suffering from engine problems, was hunting off the coast of Chile for the *Leipzig*, whose radio transmissions had been intercepted the day before. At 16.20 smoke from the *Leipzig* was sighted, but what Cradock was unaware of at that time was that the whole of Von

Below:
Von Spee's German squadron is seen here in rough seas off the coast of Chile in November 1914. *Scharnhorst, Gneisenau, Leipzig, Nurnberg* and *Dresden* proceed in line ahead. *IWM Q50992*

Spee's force was in the area. When the picture became clearer, Cradock chose to stand and fight, although his armoured cruisers mounted only 2×9.2in plus 16×6in in *Good Hope* and just 14×6in in *Monmouth*, against 8×8.2in in each of the German armoured cruisers. In the ensuing action, the Battle of Coronel, *Monmouth* and *Good Hope* were sunk, although *Glasgow* and *Otranto* escaped.

The presence of Von Spee off the west coast of Chile disrupted completely the unconvoyed merchant traffic: it was a classic use of a raiding force to prevent the flow of essential commerce; their mere presence in the area was sufficient. Nevertheless, Von Spee himself faced considerable problems, due mainly to the fact that he had no firm base to which he could return and no repair facilities should any of his ships be damaged. He was dependent upon a team of colliers to keep him supplied with fuel, and the management of those colliers was dependent upon wireless communications which were far from reliable in those days.

After his victory at Coronel, Von Spee believed he would be hunted and, despite his success, began to feel vulnerable. He resolved to take his force back to Germany after rounding Cape Horn. After coaling and provisioning, and delays caused by bad weather, it was not until 2 December that Von Spee's force rounded the Horn and entered the Atlantic. Meanwhile the British had reacted to the news of the events off Coronel.

On 4 November the Admiralty ordered Adm Jellicoe, Commander-in-Chief of the Grand Fleet, to send two of his battlecruisers to the South Atlantic with all speed.

Left:
Adm Sir Frederick Sturdee on the left, who as a Vice-Adm commanded the battlecruiser force sent to the South Atlantic. Rear-Adm Keys is on the right. *IWM Q22960*

Below:
***Invincible*, who, with *Inflexible*, was sailed from the UK to the South Atlantic to oppose Von Spee's squadron.** *Real Photos*

The advantage of sending battlecruisers was that they had the speed to catch Von Spee's cruisers and also the guns to out-range them. *Invincible* and *Inflexible* were chosen. Both these ships had been completed in 1908 and were armed with 8×12in guns in four turrets. They had been the first battlecruisers and the design sacrificed armour for speed, both being capable of over 25kt. They left the North Sea and proceeded to Devonport to coal and top up with stores, and also rectify some minor defects. Eventually the pair sailed on 11 November under the command of Vice-Adm Sir Frederick Sturdee. They proceeded south at economical speed, coaling in the Cape Verde Islands and then at the Albrolhos Rocks off the coast of Brazil. Intelligence on Von Spee's movements was sparse, although Capt John Luce, the Captain of *Glasgow*, which had rendezvoused with Sturdee at Albrolhos Rocks, was confident that the Falkland Islands were likely to be Von Spee's next target. Fortuitously, Sturdee's battlecruisers reached the Falkland Islands on 7 December.

Left:
Although in a static role in Stanley Harbour, *Canopus* fired the first shots in the Battle of the Falkland Islands. *MPL*

Right:
***Kent, Inflexible, Glasgow* and *Invincible* at 09.45 on 8 December 1914 as they raise steam before sailing to chase the German ships.**
IWM Q20895

Below:
***Scharnhorst*, Von Spee's flagship. The German armoured cruisers mounted 8×8.2in guns as their main armament — no match for the British battlecruisers.** *IWM Q22361*

Meanwhile, other forces had been gathering. After Coronel, *Canopus* had withdrawn to the Falkland Islands to act as guardship. She was positioned in the inner harbour at Port Stanley where her 12in guns could command the approaches to the harbour. Her topmasts were sent down to make her less conspicuous and some of her light guns landed. Extemporised mines were laid and lookouts were placed ashore on Sapper Hill. *Canopus* was joined by the armoured cruisers *Carnarvon*, *Kent* and *Cornwall* and light cruisers *Glasgow* and *Bristol*.

Capt Luce had recognised the threat from Von Spee correctly, for the latter had indeed decided to attack the Falkland Islands, with the principal aim of destroying the wireless station there. Thus it was that early on the morning of 8 December 1914, Von Spee ordered *Gneisenau* and *Nurnberg* to close Cape Pembroke Light whilst the remainder of the ships lay off. From *Canopus* they watched events unfold:

'Early in the morning at about 7am smoke was reported on the horizon to the south of the island. A little later a report from the lookout station said that the German Fleet had been sighted, consisting of 8 ships, making for the island. Then the fun began.

Guns crews were piped to close up and everything was got ready for action. Now to avenge *Good Hope* and *Monmouth*. One large four-funnelled cruiser and a small one with three funnels were making for the wireless station, and the remainder were steaming up slowly, little thinking what was in store for them. The two ships closed on the island to around 16,000 yards and altered course to starboard so as to bring their guns to bear on the wireless station. *Canopus* asked permission to fire, she being in a position to fire over the land, and yet could not be seen by the German ships. When we let go the 12in guns, what a surprise they must have got! They at once went full speed ahead to get out of the range. Altogether *Canopus* fired five rounds.

'This happened about 9.15am. From a position ashore, they were able to see that one of our shells hit her, and smoke was seen to be coming from the cruiser's upper deck. They were in such a state of fear that they forgot to fire at the wireless station . . .

'In the meantime our fleet had cast off the colliers from which they were coaling and got under way out of the harbour and gave chase, the Flagship leading. The men on *Canopus* had one consolation — they had saved the wireless station and town from being destroyed. Also they had the honour of firing the first shot.'

The German ships had been first sighted at 7.35 but, due to difficulties in signalling, it was not until 8.00 that Sturdee received the news. The battlecruisers were busy coaling at the time and did not have steam up. The Admiral ordered *Kent*, the only cruiser with steam up, to weigh anchor and leave harbour, to patrol off the harbour entrance. Meanwhile all was frantic activity in the remaining ships as the colliers were cast off, the last of the coal was poured down the coal chutes into the bunkers below and the stokers flashed up the boilers. It was to take some two hours before the ships would be ready to sail. Meanwhile it was at 9.15 that *Canopus* fired the first shots of the day.

The Germans were completely surprised by *Canopus*' first salvo and then shortly afterwards a lookout reported from the foretop of *Gneisenau* that he could see tripod masts behind the land. At first he was not believed, but then it was realised he was correct. Tripod masts were only fitted to British battleships and battlecruisers and when Von Spee heard the news, he knew he was up against a superior force in gunnery terms, although he probably had the speed advantage over battleships . . . if they were battleships and not battlecruisers. Von Spee ordered his ships to withdraw to the southeast at high speed.

Soon after 10am the British ships were ready to sail in pursuit. A Midshipman in *Carnarvon* describes the sight:

'So we went out, followed shortly by the battlecruisers, with the signal "General Chase" flying. It was one of the most moving moments of one's life although it was not realised at the time. It is difficult to express the exultation we felt . . . There were other features. December is, of course, high summer and I was particularly struck by the scene with the sun shining from a cloudless sky and the relatively warm, fresh breeze. It was a lovely sight with the ships steaming out . . . Shortly we were followed out by the battlecruisers. Shortly after 11am the whole fleet was at sea. Then came one of those anti-climaxes so frequent in war. Sturdee, now sure that the enemy force could not escape, slowed down to let the slower cruisers *Carnarvon* and *Cornwall* catch up. For good measure, ships' companies were sent to dinner, a wise move as the coaling had left the men exhausted, although it seemed almost sacrilege to a young Midshipman . . .'

As the British came out of harbour, the German ships were working up to full speed and watching anxiously to identify the British ships as they emerged from behind the land. When they saw the two ships with tripod masts appear, with smoke pouring from three funnels, they knew they had to be battlecruisers — no British battleships had three funnels. And every class of British battlecruiser had a speed advantage over the *Scharnhorst* and *Gneisenau*, and was armed with 12in or 13.5in guns, a minimum of eight of them. Imagine the horror when the Germans realised that they were up against two ships which could outrun and outgun them, and that their fate was almost certainly sealed unless the weather or a breakdown in one of the battlecruisers intervened. But it was a clear, calm day with many hours to go before the sun would set . . .

Sturdee signalled 'General Chase' and the battlecruisers set off in pursuit of the columns of smoke on the horizon which marked the German squadron. He knew that time was on his side and the opportunity was taken to wash off the coal dust and, as in *Carnarvon*, send the men to dinner. A stoker in *Inflexible* describes events:

'With white ensigns at the mastheads, we steamed out of harbour in pursuit of the enemy . . . We all had a meal as it was to be a long chase . . . When we were nearly in range, Action Stations were again sounded and for this action I was in the fire party amidships. About 1pm the firing started. Clouds of coal dust were everywhere and with each salvo the ship shook and rattled. We had an undercover position but every now and again we climbed up on the casing to have a look at what was going on . . . I noticed the deck was scattered with pieces of shrapnel. We had been hit somewhere and through the brilliant flashes I could see clearly that the *Scharnhorst* was firing at us.'

From the German ships, they could see that the British ships were gaining on them and that escape was impossible. Against the clear sky they could see the huge clouds of smoke which marked the battlecruisers, and then, from the base of the clouds, flashes of gunfire could be seen. It was 12.56.

Despite having the advantage of speed and gun range, it was to take some 3½ hours to sink the first German ship. Sturdee did not wish to close within the range of the Germans' 8.2in guns and therefore chose to fire from long range. However, the gunnery of the battlecruisers was also severely hampered by their own smoke. From *Carnarvon*, they watched the action:

'At 11.15 the men of *Carnarvon* were allowed on deck as we were some way astern. At 12.56 the battlecruisers opened fire. Their shots were off as they had not got the range. The enemy soon started to reply. *Carnarvon* went to dinner at 11.30. We were still some distance out of range.

'At around 12.30 however we were all back at our Action Stations, and the Admiral decided to close in . . . Once Spee had assessed the situation, he detached his smaller ships to the southward and they were pursued by *Cornwall* and light cruisers. *Carnarvon*, slow though she was, managed to get into the fight by cutting corners and just about 3pm the battlecruisers and *Carnarvon* recommenced the action. Within a few minutes the *Scharnhorst*, despite first-class gunnery, started to list and then sank. It has been said that a ship should have been left to pick up survivors but the battle was still in full spate . . . I recorded that at 4.16pm the *Scharnhorst* started to go after putting up a fine fight and some very fine shooting . . .'

In fact Sturdee had opened the action at 16,000yd. At first the British gunnery was not very accurate and no hits were scored. However, at about 1.30pm, Von Spee took the initiative and turned the two German armoured cruisers towards the British ships, whilst his light cruisers made off to the south. Although Sturdee turned his ships to parallel the Germans, Von Spee was able to close the range sufficiently to obtain hits on *Inflexible*, but no great damage was done. At 2pm the Germans turned away again. The battle reached something of an impasse with the Germans unable to score effective hits and the British gunnery still rather ragged and still being affected by smoke. In an effort to get into a more favourable position to be clear of his own smoke, Sturdee allowed the range to open and firing ceased for a while.

By 3pm Sturdee had closed the range once more. Von Spee, once again, turned to close in an attempt to inflict damage on the battlecruisers. The German gunnery was good and hits were indeed obtained but 8.2in guns were unable to cause mortal damage, even the comparatively light armour of the battlecruisers being sufficient to prevent the shells from penetrating. At last, the British ships started hitting *Scharnhorst* and their 12in shells penetrated the armour of the cruiser, doing tremendous damage. *Scharnhorst* was torn to pieces, although her men manned the guns and kept firing to the last, until she turned over and sank. There were no survivors.

The crew of *Gneisenau* could only watch *Scharnhorst* being annihilated, knowing they would also suffer the same fate. However, the crew fought on magnificently as their ship was reduced to a shambles around them. By 5.15 she was little more than a hulk, the decks above the armour unrecognisable, her engines and boilers out of action. She sank at 6.02pm. From *Inflexible* they watched the final moments:

'Soon after, there was a lull in the firing and then we heard that the *Scharnhorst* had gone down with all hands . . . We had ceased firing by then, only to find that the other ship (*Gneisenau*) had fired. At last her firing ceased altogether. The upper deck of our ship was crowded with men watching the last of the *Gneisenau*. We saw her suddenly turn on her side and slowly vanish, leaving hundreds of men in the icy water.'

Carnarvon also watched the final moments:
'I was in a bit of a funk, which was at least honest. However, there came the moment when we were all called on deck . . . What a difference between the morning and the evening. We had gone out with the sun shining down upon us, flags flying, men cheering, twenty odd knots, chasing five columns of smoke on the horizon. In the last phase, the grey overcast, lowering sky, the bitter cold wind blowing through the shrouds and a thin sleet beginning to fall. Three ships lying in a ring around what had been that morning the best of the German Navy . . .

'Boats were called away to rescue survivors [from *Gneisenau*]. As I was the Midshipman of the cutter, which was the seaboat, we were the first to be ready as the remainder were lashed down and filled with water to avoid fire risks . . . The sight of the ship slowly turning over was a sad sight and obviously had a profound effect upon me. The water was very near freezing point and their sailors were, in any case, extremely exhausted . . . I recall what a grisly sight it was. There was now something of a sea running which made it terribly difficult to haul the men out of the water. It was only done at the cost of considerable effort from our own crew . . .

'One of the survivors was of particular interest as he was the signalman . . . who, from the *Gneisenau*, had reported that he could see tripod masts . . .'

Meanwhile the British cruisers had given chase to their German counterparts. By the end of the evening *Nurnberg* and *Leipzig* had been sunk. Only *Dresden* escaped, and Von Spee's East Asiatic Squadron was no more.

Left:
***Invincible* building up to maximum speed during the chase after the Germans. The vast amounts of smoke pouring from her three funnels were to cause problems for the British gunners.** *P. Liddle/Cdr J. G. D. Ouvry*

Above:
The sight from the foretops of the British ships — the plumes of smoke made by the German squadron taint the horizon. It was a long chase, but the British battlecruisers had a considerable speed advantage. *IWM Q20894*

Right:
***Inflexible* opening fire during the Falkland Islands engagement, as seen from *Invincible*.** *IWM Q20891*

The Dardanelles

1 The Beginning

The Dardanelles and Gallipoli are names which tend to conjure up visions of men fighting in desperate conditions on land. However, the Dardanelles campaign started as a purely naval venture and involved a considerable number of battleships. Whilst the actions in the North Sea and off the Falklands lasted only a matter of hours, battleships were to be involved in the Dardanelles campaign for weeks, indeed months, and they were to go into action many times and suffer significant losses. It is probably the longest sustained operation undertaken by battleships and is thus worthy of consideration at some length.

In January 1915 it was decided to prepare a naval expedition whose objective was to bombard and take the Gallipoli peninsula, with Constantinople as the ultimate goal. This meant forcing a passage of the Dardanelles, the narrow strait which connects the Aegean Sea with the Sea of Marmora and thence to the Bosphorus, upon which lies what is now Istanbul, then called Constantinople. The Dardanelles were well protected, with forts on either side of the entrance; further up towards the Narrows were lines of mines, themselves covered by the guns of further forts on the shore. In addition, the Turks had mobile howitzers which also covered the approaches to the minefields.

The fleet which was to attempt to force a passage through these defences consisted of up to 18 battleships and battlecruisers, British and French, supported by destroyers and minesweepers. So as not to weaken the Grand Fleet in the North Sea, most of the battleships sent to the Dardanelles were the older ships, mainly of the 'Majestic', 'Canopus', 'Formidable' and 'London' classes. Most of them were armed with 4×12in in twin turrets, 12×6in and a number of 12pdrs, and were capable of speeds of about 17-18kt. They all had triple expansion engines and were coal burners. The one exception which was to arrive on the scene was the brand-new *Queen Elizabeth* with 8×15in, steam turbines and oil-fired boilers.

Operations commenced on the 19 February 1915. The first requirement was to silence the forts in the vicinity of Cape Helles and Kum Kale, but the ships quickly found that this was no easy task. It was no good just hitting the forts: the actual guns within the forts had to be hit and the ship's gunnery was just not that precise. Also it was quickly found that the return fire of the forts and the mobile howitzers was disconcertingly accurate. A Midshipman in *Agamemnon* describes the plan for 21 February:

'The following is the scheme for today's bombardment:

'General plan — *Vengeance* and *Cornwallis*, *Suffren* and *Charlemagne,* working in pairs, will run in to 3,000 yards and engage Forts No 1 and 4 with their secondary armament aiming to destroy each individual gun by direct hit.

— In support *Agamemnon*, *Q.E.*, *Irresistible* and *Gaulois* firing at long range deliberately at Forts 1, 3, 4 & 6 respectively with the object of preventing the defenders from manning their guns.

'Ships attacking at close range:

— ships of each pair are to be at such distance apart that the one can support the other in turn. Each one will commence by order of the V.A.

Below:
***Lord Nelson* — the battleships sent to the Dardanelles were mainly pre-Dreadnoughts; *Lord Nelson* sported 4×12in and 10×9.2in guns.** *Real Photos*

commanding. The second ship of a pair must not turn in the wake of the leader.

'Supporting ships:

— these ships will anchor in positions indicated and range on their forts, reporting when range is obtained. Once firing has commenced, supporting ships will open a slow fire on forts. *Agamemnon* and *Q.E.* will use ¾ charges from their 9.2in and 18in [sic] guns. *Irresistible* 3 charges from 12in guns. *Agamemnon* and *Irresistible* are to cease fire when the forts are being closely engaged by the inshore ships.

'Unfortunately when we arrived in sight of the Dardanelles, we turned round and went back to Tenedos as the wind was so strong we decided not to carry on with the bombardment. There was rather a mist on the horizon too which would have made it very difficult to spot.'

There were still a lot of lessons to be learned — it was early days. It can be said at this stage that the ships were only operating during 'working hours'. They shelled the forts by day but at night retired to anchorages at Imbros or Tenedos, where there was, at this time, no threat of enemy attack. The next extract shows how the realities of the Dardanelles were learned rapidly in *Agamemnon* on 25 February:

'When the shells began pitching close to us I was standing on the fo'c'sle with a crowd of men. However, the Commander soon came along and cleared the fo'c'sle. I went back to the fore-turret. However, soon an order came for the turret's crew to fall in aft. I went into the port battery and the Commander then tried to start us on painting the side of the ship . . . I have never heard anything so utterly ridiculous and suicidal. The first shells soon put a stop to that. This hit a leg of the main derrick and splinters from it killed Petty Officer Worthington on the signal bridge. At the time I did not realise that the ship had been properly hit. The noise of our own guns firing drowned everything. I thought that some rigging had merely been knocked down. I went out of the port battery and stood on the break of the fo'c'sle . . . Almost immediately afterwards a shell came through the flying deck port side and burst . . . Splinters were scattered everywhere into the innermost corners of the port battery . . . By this time we were ordered to weigh anchor. The Chief Stoker had turned on the steam to the cable holder and the cable began to come in slowly . . . Shells were falling all around, short and just over, and while we were weighing the ship was hit 4 times.'

The example above of a Commander trying to have the ship painted whilst bombarding the forts shows the naivety common in the first week or so. The ships quickly settled into a routine. For the gun crews, it could mean long, stuffy days in confined conditions. A typical crew for a twin 12in turret consisted of a Lieutenant, two Midshipmen and 16 men actually in the turret with an armourer and four seamen in the gunbay, a Petty Officer plus eight in the magazine and handling room and a further Petty Officer plus 10 in the shell room. Spotting for gunnery was usually done from the foretop, although some use was made of aircraft. These had their limitations due to the rudimentary communications available and the lack of endurance, but they did prove useful for general reconnaissance. Gunnery was found to be too inaccurate whilst under way so the tactic of ships anchoring off and providing covering fire to drive the Turkish gun crews from their positions, whilst other ships closed in to finish off the guns

Left:
Tongues of orange flame lick from the gun barrels as *Agamemnon* engages targets ashore with her 9.2in. The 9.2in guns were often used in preference to the 12in. *MPL*

Below:
French battleships in action in the Dardanelles during 1915. Four French battleships took part in the campaign. *IWM Q53526*

Right:
The type of damage caused by Turkish howitzers — *Agamemnon's* quarterdeck after such a hit creates a frame for these two Naval and Marine Officers. *MPL*

themselves with short-range fire, started to produce results. Nevertheless, the British and French ships found that it was the mobile howitzers ashore which were the greatest problem, as they were difficult to locate in the first place and could shift position if threatened.

A main base at Mudros was built up with the battleships using anchorages at Imbros and Tenedos as forward bases. It was at these anchorages that the ships coaled and ammunitioned. Coaling was required about once per week on average; ammunitioning might be required more often, depending on expenditure. For example, on one day in March *Lord Nelson* fired 145 rounds of 9.2in and 80 rounds of 12in. The rate of fire was not high — typically one round per minute — but even allowing for the fact that not infrequently firing was stopped for lunch and tea (!), over a long day ammunition expenditure could be considerable.

However, as mentioned earlier, at this time there was no threat to the battleships at the anchorages at Imbros and Tenedos and it was still the routine to continue with such things as Divisions, and Church on Sundays. Indeed, operations were often cancelled due to poor weather and an almost peacetime harbour routine could be maintained on some days. There was still time for sailing, deck hockey, walks ashore and other leisure activities.

As the outer forts were slowly overcome, demolition parties were sent ashore to destroy the remaining guns. *Agamemnon* supported such operations on 4 March:

'We got underway at 7.15 and steamed towards the entrance. We had been ordered to cover the landing parties near Fort No 6 . . . When we arrived, *Cornwallis*, *Irresistible*, *Majestic* and *Ocean* and several destroyers were already in the entrance. Our two cutters were sent to one of the destroyers to land some Marines which they had onboard. . . . We steamed round in the entrance firing 12pdrs the while at Yenisher . . . A suspicious object was seen in the water and we fired at and it sank. I think it was afterwards discovered to have been a submarine mooring buoy.

'We continued to fire at Yenisher with our 12pdrs. At 1 o'clock we fired 3 rounds from S2 turret at a battery behind and to the left of No 4. At 2.15 we opened a heavy fire with our 12pdrs at the left edge of the village where several Turks were seen. We made them hop alright! I could see them through my telescope running for all they were worth. We fired from P2 turret at the battery at the same time. . . . I'm afraid our demolition parties which were trying to get to No 4 fort were having a very bad time. All afternoon and up until dark we could see them running along the beach for short distances and then dropping flat. Men could be distinguished crawling on hands and knees . . . Our cutters' crews coming back were stating that there had been a considerable number of casualties.'

Eventually such methods were successful. Having dealt with the problem of the outer forts, although the mobile howitzers were still a considerable problem, attention was turned to the defences of the Narrows. The difficulty was that the forts protecting the Narrows had to be knocked out before minesweepers could tackle the minefields. However, the battleships would be hard-pressed to deal effectively with the forts unless they could get close, and this they were prevented from doing by the same minefields. In order to force the Narrows, considerable losses of ships were anticipated by the high command, and the risks involved were quickly realised by the officers and men in the ships once they came to grips with the problem.

Operations against the Narrows were commenced in early March. Attempts were made to shell the Narrows forts from long range with only limited success and the minesweepers also tried to start sweeping the mines. The minesweepers attempted to sweep by night but searchlights ashore made

Right:
Sticks! There was still time for recreation during breaks from bombardment when the ships retired clear of the Dardanelles. Officers are seen keeping fit playing deck hockey.
P. Bosworth

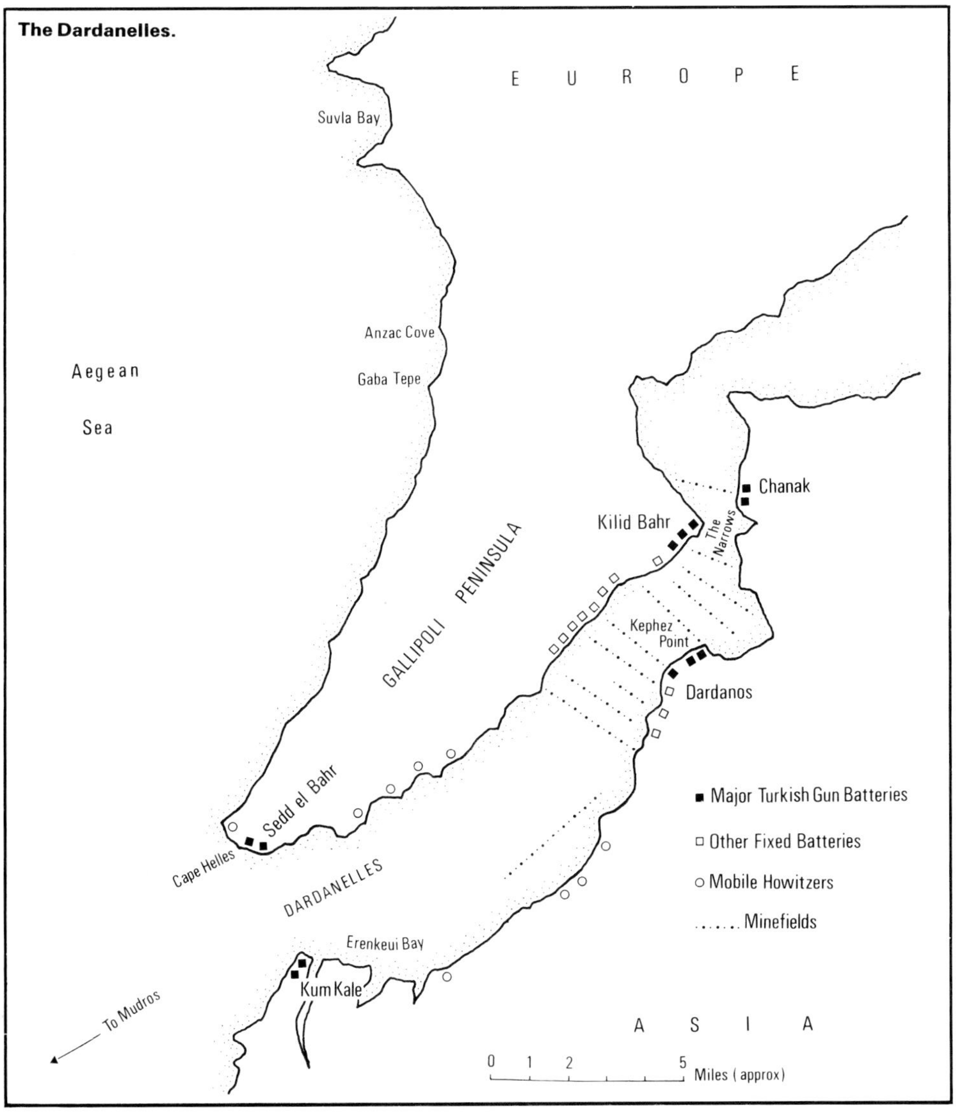

certain that they came under the attention of the guns. In addition, free-floating mines were becoming a hazard. The outflowing current in the Dardanelles was ideal for the Turks to float mines down from the area of the Narrows. Nevertheless the guns ashore remained the greatest hazard. A Stoker from *Lord Nelson* describes a typical day during this phase of the operations:

'Underway for Dardanelles at 9am. We have been told off for Forts No 13 and 19. No 19 is at Chanak at entrance to Narrows . . . French ships to cover us from moving batteries. "Aggy" led way in at 10am. We followed 6 cables astern. "Aggy" commenced off No 10. We fired the first round from the fore-turret and got a lovely hit. We were then within range of Chanak Forts which mount 2×14in and 9×11in guns. Continued to engage both forts and done considerable damage. Fleet did not think we were coming out again. "Aggy" had to retire for half an hour through being hit. The experience of being under fire is not so bad as I thought it was going to be. I was stationed on the fore bridge when the shells were dropping all around. I thought it was about time that I got under cover and hadn't been down very long before one arrived on the bridge about 10 feet from where I had been standing. We were hit several times, one hitting between P1 and P2 turrets and one 12 feet below the waterline, causing a leak in a bunker. This was not discovered

Below:
Mudros Harbour became the main base for operations in the Dardanelles. This photograph was taken some time after the big build-up in troops, as indicated by the large huddle of tents in the foreground. The battleships are anchored on the left of the picture. *IWM Q13820*

Right:
***Agamemnon* in action in the Dardanelles, with shells from the Turkish batteries falling around her.** *MPL*

Below:
***Triumph*, another of the Dardanelles force, had the unusual armament of 4×10in and 14×7.5in guns, having been ordered originally for Chile.**
IWM Q40365

until we were anchored and the ship was getting a list to port. Divers were sent down to ascertain the amount of damage. It was found not to be serious enough to go into dock.'

Long-range fire against the forts did obtain hits on the forts themselves, on occasions from ranges as great as 21,000yd, but spotting remained difficult and the guns themselves within the forts were not silenced.

On 8 March an event took place which was eventually to change the whole course of events in the Dardanelles. By night the 380-ton Turkish ship *Nusrat* laid a line of 20 mines in the northeastern part of Eren Keui Bay, in the area used by the bombarding battleships. This line was laid up-and-down current, unlike all the other lines which were across the current, and was some 3,000yd long. The mines were laid at such depth that only battleships would make contact with them.

By luck, and also due to the fact that operations were curtailed somewhat over the next few days, the minefield remained undiscovered. There was a definite lull at this time as an excerpt of *Lord Nelson's* movements, or lack of them, indicates:

'10 Mar Mudros. Still nothing doing. Cleaning ship a bit now. *Queen Elizabeth, Inflexible*, "Aggy" and Russian cruiser *Askold* arrived here. Transports still coming in.
11 Mar Mudros. Same old thing — waiting. *Inflexible* left, presumably for England. *Hussar* arrived here, also more transports.
12 Mar Mudros. Still waiting. Cannot get any news. Mail arrived.
13 Mar Mudros. Same old thing.'

Eventually, on 18 March, a large operation, using all the available battleships was mounted. *Vengeance, Irresistible, Albion, Ocean, Majestic, Prince George, Queen Elizabeth, Agamemnon, Lord Nelson, Inflexible, Swiftsure* and *Triumph* formed the British force, and *Gaulois, Charlemagne, Bouvet* and *Suffren* represented the French. The basic plan called for an initial group of ships to shell the forts from a long range to force the gun crews to retire, after which a second line of ships would move in closer to knock out the guns. However, it was the howitzers and the *Nusrat's* minefield which were to dictate the course of events. A Midshipman in *Agamemnon* describes the start of the day:

'We began to get under way at 8.20. At 8.30 we were to be prepared for immediate action. I managed to get a fairly good breakfast first. Packets of sandwiches were made up for everyone. We had Divisions and Prayers at 9.10. Afterwards Guns held forth on the plan of operations. *Q.E.* with *Prince George, Inflexible, Lord Nelson* and ourselves were to go up to a range of about 14,000yd and deliberately bombard the forts at the Narrows. Fort 13 fell to our share amongst others. When the forts seemed to be sufficiently silenced, 4 French ships were to pass through the First Division which formed Line A and go to 8,000yd from the forts. The outside ships had been detailed to locate and destroy torpedo tubes which are thought to exist on the European and Asiatic shores. The French ships were to be relieved by *Vengeance, Irresistible, Albion* and *Ocean* off the entrance.'

Ashore the Turks could see the force gathering and prepared accordingly. A Turkish battery officer describes the scene: 'We saw a mass of ships such as we had never seen before. We were amazed and, realising that that day we should be faced with an out-and-out conflict. We completed our supplies and prepared for the attack. We knew each ship well from the lists we possessed.'

The plan started reasonably well, as the account from *Agamemnon* continues:

'Action sounded at 10.30. At 11 we came under the fire of howitzers and field guns and shortly afterwards we opened on No 13 with the fore turret after first firing a few rounds from P1. The shooting of the fore-turret was very good and the Captain complimented the turret on it. *Prince George* and *Q.E.* were nearest the European side, then came us, *Lord Nelson* and then *Inflexible*. At 12.15 the French squadron commenced to close in and pass through Line A. At the same time we continued our deliberate fire at the forts. There was no reply from the forts but single howitzer batteries were firing.'

The fire against the forts appeared to be having the desired effect so, as planned, the French ships closed in to shorter range. By this time it was just after noon and the pace of the action increased considerably. As the French ships eventually turned away, *Bouvet* became the first major casualty.
[Note that some observers believe it was shellfire which caused the damage whilst others believe it was a mine. In fact, it would appear that she was the first victim of the *Nusrat's* minefield.] From ashore the Turks watched:

'The battle developed with considerable violence and by noon the French ships in the second line advanced through the first line and opened a tremendous bombardment. The batteries replied effectively. Under this fire the *Bouvet* started to withdraw, then a cloud of red and black smoke arose from the ship. She may have struck a mine. Immediately after this there was a much more violent explosion. We believe that a shell from Mejidiye had blown up the magazine. The ship heeled over at once and her crew poured into the sea. On both sides fire ceased. The destroyers in the rear hastened to save the crew.'

From *Agamemnon* they also watched:

'At 1.55, the French ships returning and the third Sub going into relieve them, *Bouvet* was struck by a shell in the magazine and sank at 2pm in 1½ minutes. No explosion was seen and she was coming down the Straits at full speed. She began to list more and more to starboard and eventually turned turtle, going down stern first. Destroyers and picket boats went to save life but there were only a few survivors. They stated that the magazine blew up.'

And from *Gaulois:*

'*Bouvet* turned in order to place herself behind the *Suffren* and she hit a mine. A small sheet of flame and yellowish smoke appeared from the starboard 27cm gun turret. For a few seconds she continued her course and then gently without any explosion listed to starboard. The Commander had clearly had no time to stop the engines. As we saw her sinking lower and lower without losing speed. . . *Bouvet* turned completely over to starboard and disappeared.'

From then on, events turned against the British and French. Midshipman Banks

Above:
***Bouvet* started to withdraw, then a cloud of red and black smoke arose from the ship . . .'**
IWM SP682

describes the fate of *Inflexible* and *Irresistible* who were badly damaged within minutes of each other:

'4.11 *Inflexible* struck a floating mine and proceeded out of action with a slight list to starboard.

4.13 *Irresistible* also struck a floating mine and took a list to starboard and was down by the bow. She made in towards the Asiatic shore. . .

4.30 Destroyers closed with *Irresistible*, the destroyer leader going alongside and taking off the crew. *Irresistible* was under fire from concealed howitzers all the time. . .'

For a stoker on *Inflexible* it was all too close for comfort:

'We could hear the thunder of guns above the engines, together with the shuddering and blast showering dust everywhere, though we had no knowledge, of course, of how things were going. Then there was a thud which shook the ship. It was followed by silence from our guns. Then we could hear the thud of the tread of feet on the deck. Meanwhile all the electric lights had failed and we lit our secondary oil lamps which gave only a dim light. The engine room telegraph rang "Half Ahead" and remained at that long enough for us to surmise that we were either through the Narrows or we were retiring. The matter was confirmed by a well-meaning voice shouting down the lift shaft "Abandon Ship Stations!".

'We were still unaware of what had happened until the Engineer of the Watch received from the bridge a message stating that we had been mined forward and, if possible, we would make for Mudros and beach ourselves in shallow water. Of the watches down below, one watch should go to its Abandon Ship Station, the other should remain down below to the last moment. Each man tossed up with his opposite

Below:
A 14pdr gun on *Triumph* knocked out by an 8in shell from a Turkish fort on 18 March 1915.
P. Liddle/P. May

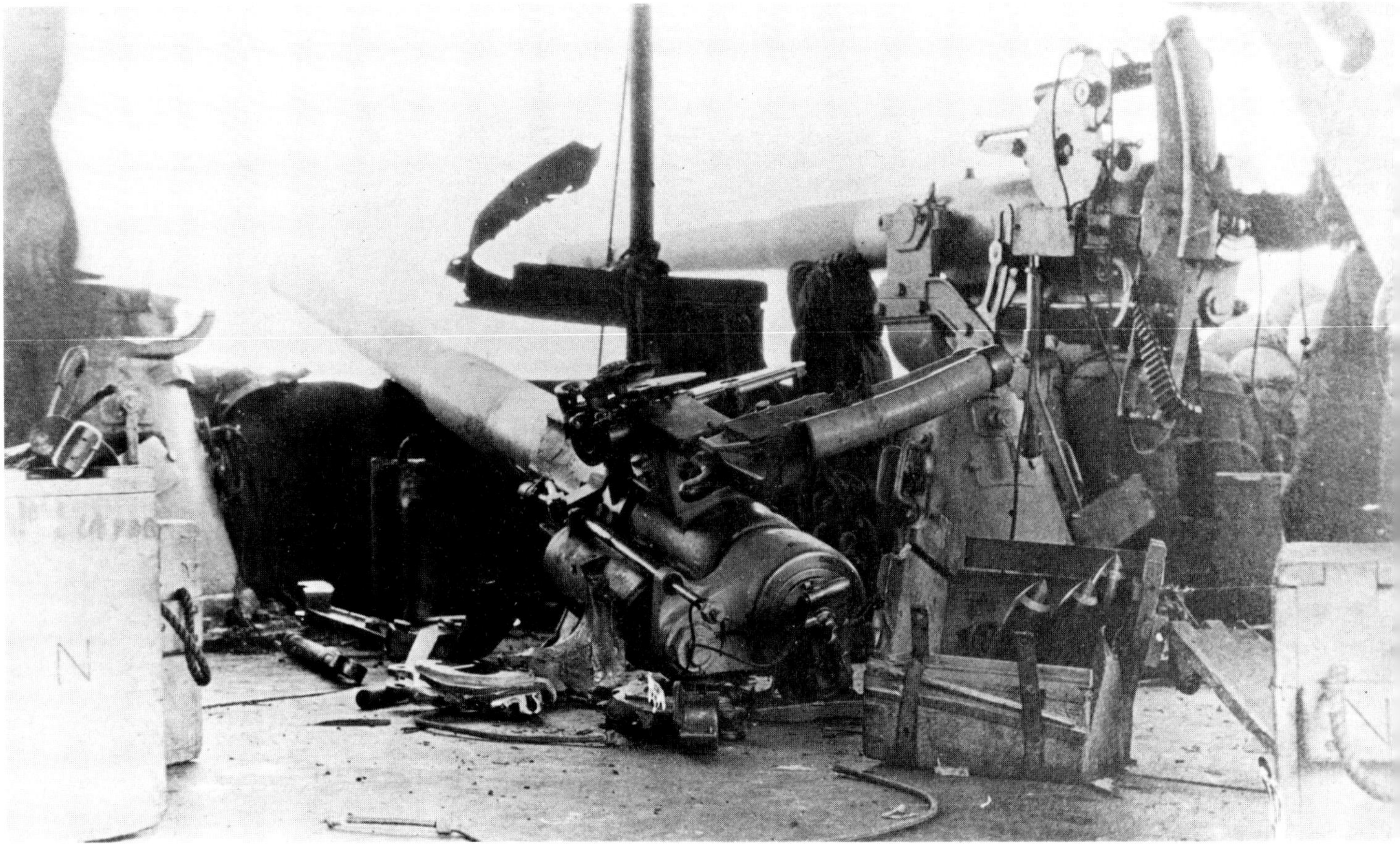

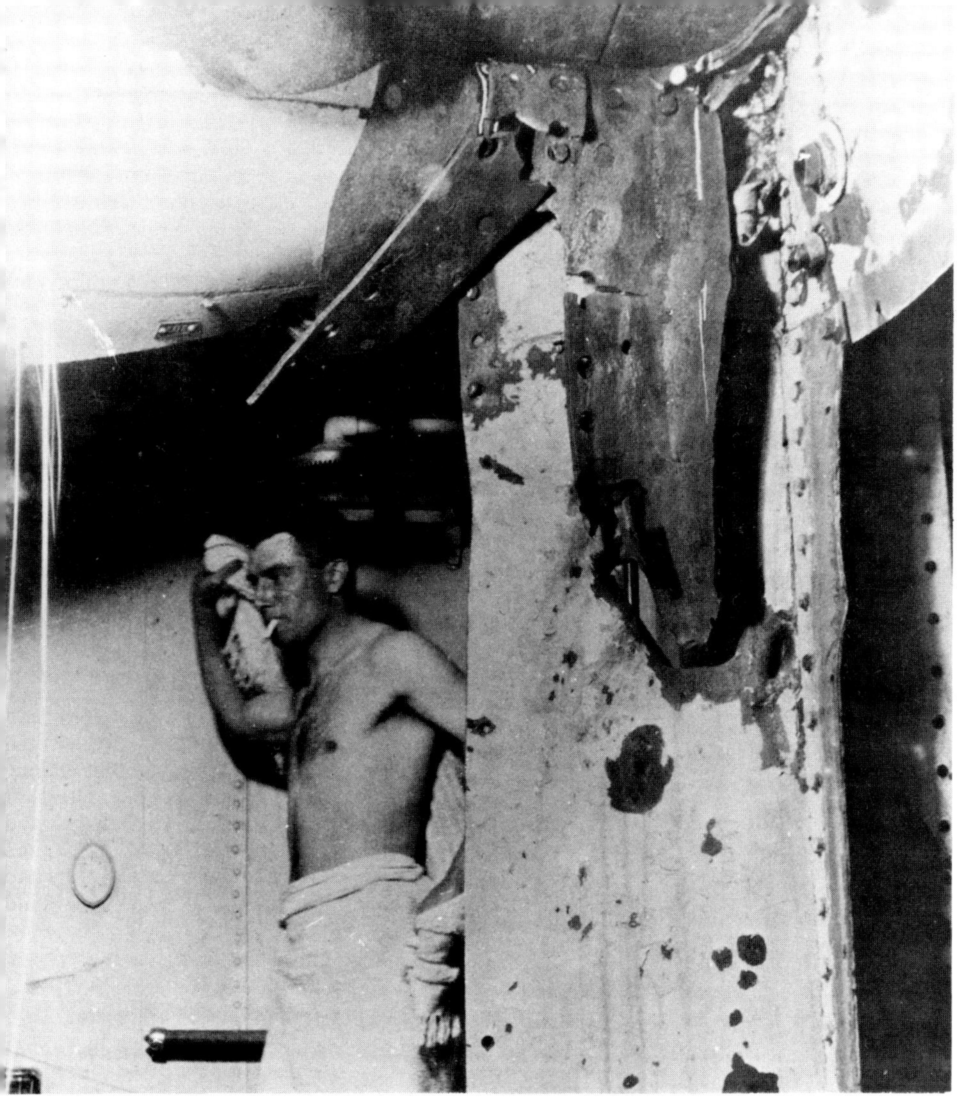

in the forward submerged flat and 40 hands were killed who were in there at the time.'

Later, as the survivors were waiting to be taken off:
'Our people were waiting quietly for the destroyers to take them off. One shot fell right in the middle of the quarterdeck and must have wrought awful havoc, but there was no panic. None tried to get away before ordered to and the destroyers were wonderfully handled and saved the situation completely.'

There was more to come:
' "General Recall" was hoisted at 6.05. At 6.30 *Ocean* was struck on the starboard side by a torpedo or mine. The Engine Room was flooded and the helm was jammed hard a-port so she could only move in circles. She listed about 15 degrees to starboard. Destroyers went alongside and took off all her crew. 8.15 the remainder anchored off Tenedos.

'The *Ocean* had the next billet to the number. I won against my friend. . . we were urged to get a move on as the ship was developing a list and the bows were becoming deeper in the water also. Up we went on the various ladders groping our way as it was total darkness. We knew our way so the only problem was the watertight doors which led to the messdeck. After knocking off the clips we couldn't push it open and no wonder as we should have been pulling it! Once outside we were amazed at the shambles. First of all we could see that the mess tables and chairs had been taken for buoyancy by the men who had had to go overboard. 29 had been killed in the forward torpedo flat by the explosion of the mine, and damage by another mine had flooded the food store thus compensating for the list we were experiencing to starboard. Wounded were being placed into the boats and I particularly remember the Master at Arms sitting in a cutter smoking his pipe and making light of his injuries.'

Left:
A gaping hole above the officers' bathroom in *Triumph*. The shell penetrated the quarterdeck on 18 March 1915 and burst inside. *P. Liddle/ P. May*

Below:
The engine room of a battleship. The men in the engine rooms and boiler rooms were very cut off when in action; they could hear the thunder of the guns but had no real idea what was happening above them. *IWM Q17931*

Others were watching as *Irresistible* was first badly hit by gunfire and then mined:
3pm the *Irresistible* caught it properly. Her foretop was smashed in, all but one in the top being killed. Soon afterwards she was hit on the forebridge, setting it on fire. It was a sight to see her blazing away firing her 12in guns at the same time. That was not the worst that she was to get, for she was mined

Irresistible and it was thought that the spot must have been mined, for the *Ocean* was seen to have a heavy list and was flying distress signals. And then there was a splendid piece of work done by the destroyer flotilla. They dashed up through a proper shower of shells and took everyone off her. . . We had, of course, been shelling the forts Nos 13 and 17 as will be seen by the expenditure: ie 58 rounds 12in, 158 rounds 9.2in common and lyddite. The magazine went up at 17 Fort. "Lizzie" was doing a bit of good work to Chanak by dropping a few 15in into it, and it was blazing merrily. The "Aggy" had a lovely hole in her funnel knocking nearly half of it away, and she also had a fire on the flying deck.

'The engagement lasted nine hours and it was quite long enough. We had a very narrow escape from being mined as we saw one about 10 yards off our bow.'

Queen Elizabeth appears to have had the final say as seen by the Turks ashore:
'We opened fire but again firing ceased. Towards sunset the battle slackened and the ships withdrew. Just then the *Queen Elizabeth*, lying off Hissarlik, opened fire on my battery. She was too far away for us to reply and we withdrew the gun crews without any damage being done.'

So the ships withdrew to count the cost. *Bouvet* sank in the middle of the action. *Irresistible*, who had already had her after turret put out of action before she hit a mine, was to be taken in tow by *Ocean* who herself hit a mine; both ships were abandoned and sank unseen in the night. *Inflexible* survived hitting a mine but had eventually to withdraw to Malta, as will be described shortly. *Gaulois* was badly damaged and had to be beached to prevent her sinking. *Agamemnon, Charlemagne* and *Albion* were also damaged by shellfire. The minesweepers made a final attempt by night to tackle the minefields, but quickly abandoned it in the face of gunfire from the forts which the day's action had failed to silence.

The most badly damaged ship to survive was *Inflexible*. She was first badly hit at 12.20 when a hit on the centre leg of the tripod foremast set fire to the deckhouse and forebridge. At 12.23 she was hit on one of her turrets, plus three other less serious hits. At 12.27 communications were lost to the foretop. Shortly afterwards a shell burst above the foretop killing or wounding all the men in it and causing a fire. She withdrew from the action temporarily to fight the fire and remove the wounded.

Inflexible returned to the action shortly after 2.30pm. At 3.45 a violent concussion was felt externally either from a mine or a shell, but the hull was not penetrated. However, at 4.00 she definitely hit a mine: the ship was lifted bodily and then began to list to starboard. All the lights went out and ventilation stopped, but the engines and boilers remained intact. The submerged torpedo flat was flooded and the ship started going down by the head, taking in some 2,000 tons of water.

At 6.00pm she anchored at Tenedos where an inspection revealed a hole 30ft square in the side and up to 20 compartments flooded. She was moved into shallow water and temporary repairs were commenced. Thence she moved to Mudros where a cofferdam (a watertight enclosure pumped dry for repair work) was fitted over the hole. She then proceeded slowly towards Malta but had to be towed stern first for the final few hours as the cofferdam was working loose. After further repairs in Malta, she proceeded to Gibraltar where it took a further three months to restore her to full fighting condition, whereupon she rejoined the Grand Fleet.

Below:
Inflexible in dock in Gibraltar after being mined in the Dardanelles. She was first towed to Malta for temporary repairs and then spent three months in Gibraltar before rejoining the Grand Fleet. *MPL*

3 In Support

The naval attempt to force the Dardanelles was abandoned and plans made to land troops on the Gallipoli peninsula. Meanwhile the battleships supported by destroyers and minesweepers continued to keep the entrance to the Dardanelles under Allied control by regular sorties to shell the outermost defences. *London* had arrived to join the force and one of her Midshipmen describes the routine:

'The routine to which we are working is roughly as follows. During the night a cordon of destroyers is kept between Cape Helles and Kum Kale, while two battleships from the 2nd or 3rd Division patrol a little further out. At daylight these destroyers withdraw and their place is taken by the sweepers escorted by a few destroyers and supported by the two battleships who have been on patrol during the night. The sweepers proceed to sweep the allotted area while the battleships engage any forts which open fire on any sweepers, and generally harry the Turks as much as possible. At dusk the sweepers are relieved by the destroyers and the battleships resume their patrol. Every other evening, the battleships are relieved by another pair and they return to Tendos, anchoring on the northern or southern side, depending upon the wind, completing with fuel, ammunition, repairing damage, etc. Periodically they return to Mudros for rest. The idea is to keep the approaches to the Narrows clear of mines and, at the same time, prevent the Turks from strengthening their defences.'

He goes on to describe a typical day in early April:

'At 8am "Action" sounded off and we entered the Dardanelles, preceded by the minesweepers and destroyers, followed by the *Prince of Wales*. The forts on the European side at once opened fire at the small craft, hitting them several times. We replied with our port 6in guns and they ceased fire. It seemed that the Cape Helles lighthouse was being used as an observation post for these forts, so we turned our attention to this edifice. We opened fire on it with our 6in, hitting repeatedly but failing to bring the tower down. By this time the Turks had plucked up courage and opened fire on us from some concealed batteries on the south side, and from howitzers on the north side. These latter scored the only hit of the action. A shell struck the rubbing strake about 10ft aft of Y3 casemate but failed to explode. Owing to the extreme steep angle of descent of howitzer shells, it did not come aboard. One of our aeroplanes now appeared and commenced spotting for us with the result that we finally started hitting one of the Asiatic batteries which promptly ceased fire. We also ceased fire and sent our hands to dinner by watches, retiring somewhat towards the entrance.

'Towards the end of the dinner hour an area close to us on the north shore was noticed to be the centre of some activity. Two 12pdrs from the port battery manned by scratch crews opened fire on it. The shooting was remarkably accurate and the activity quickly subsided. We then went to General Quarters and prepared to give the *coup de grâce* (to the lighthouse).

'Having closed to 2,000yd, we opened fire with the fore-turret. The first two shots misfired and then on the third attempt the

Below:
***Agamemnon* at anchor with her torpedo nets out. The battleships spent a lot of time at anchor in lulls between activity; with the arrival of German submarines in the area, the threat from torpedoes became significant. The nets were suspended beneath the water from the booms protruding from the ship's side.**
IWM SP67

Right:
Getting in ammunition: ammunitioning was a constant chore, along with coaling. On returning to the anchorage after bombardment, there was often no break from activity — ammunitioning and coaling had to be completed as soon as possible. *P. Liddle/ Capt C. A. G. Nichols*

Below:
The islands of Imbros and Tenedos provided an opportunity to go ashore for relaxation. Here a party is being landed by boat on Imbros. *IWM Q13405*

gun finally went off and fired. The second round hit the tower squarely. The aeroplane had to return to base for petrol so for the remainder of the day we only fired occasional rounds from the 6in. At 6pm we secured and withdrew for the night.'

There was still time, however, for relaxation:

'9 APR A DAY AT TENEDOS
Quiet day. In the afternoon the Gunroom and a few of the Wardroom officers sailed to Tenedos for a picnic. We bathed and romped about and I investigated the flora and fauna of the island, and succeeded in catching two tiny terrapin.'

Many of the accounts of this period were written by Midshipmen. Indeed, Midshipmen were required to keep journals which were inspected regularly by their Captains, and these journals have survived to this day. Despite the operations going on, the normal requirements of naval routine continued and the Midshipmen played their part:

'Most of the Midshipmen changed their duties. From having the second cutter and keeping watch, I became "Tanky" or Navigator's assistant. Midshipmen's general duties, as distinct from their stations for Action, Night Defence, Coaling, Fire, Collision, etc, roughly speaking are as follows: Two picket boats and two cutters are sea boats which are continually in use with a Snotty permanently in charge of them, responsible for their condition and cleanliness, as well as for their actual running. The launch, pinnace and whaler have someone told off from the watchkeepers to run them. The Gunnery Lieutenant, the Torpedo Lieutenant and the Navigator each have a Midshipman as an assistant. Four Midshipmen are employed encoding and decoding signals . . . Four Midshipmen are under instruction in the engine room, or at least they should be, but at present all the remainder are upper deck watchkeepers.'

On 25 April troops were landed in large numbers on the southwestern tip of the Gallipoli peninsula. A vast number of ships were involved, including the battleships. In *Implacable*:

Left:
A long way from England's green and pleasant land: three Midshipmen from *Lord Nelson* going ashore for a picnic, complete with basket and kettle. *P. Liddle/Vice-Adm E. Langley Cook*

Below:
A Midshipman's gunroom which was the Midshipman's mess, presided over by a sub-lieutenant. Midshipmen have provided many of the accounts of World War 1 as they were required to write journals whilst on board. *IWM 18596*

'24 Apr. Entered Tenedos in the forenoon. Drew our boats from transport and also 4 picket boats from other ships. Prepared the boats and provisioned them. In the evening 700 troops of the covering force came aboard from the trawlers who had fetched them from the transports. The crews manned the transports' boats and secured astern of the ship. At about 10pm the ship proceeded to sea for the Dardanelles towing us in the boats astern. All the Fleet followed.

25 Apr. As soon as it was daylight, the boats came alongside the ship and we took the 700 aboard the boats and, in company with the ship, we closed in to the beach, the ship firing hard to cover us. Then the boats made a dash and landed the first 700 men with hardly a casualty owing to good firing from the ship. Several landings were effected along the coast but with not so much luck as us.'

In *Lord Nelson*:
'25 Apr . . . Up before the lark this morning. Weighed and proceeded to Dardanelles at 3.15am to cover the landings of our troops. Commenced at daybreak with a heavy bombardment at Sedal Bahr. The fort had been smashed up previously but they must have got some more guns there during the spell we gave them. Had also entrenched themselves well around the village. It was a very hard nut to crack. Our troops got a footing under our fire. They were fighting before they got anywhere near the shore. . . Expenditure of ammunition will show that we were not idle — 35 12in, 84 9.2in and 1,124 12pdr. We finished firing at 8.30pm, being about 16 hours in action.
'26 Apr . . . Was at Action Stations just before daylight ready to give the Turk another warm time of it. I hear that our troops had a rough time of it last night. The Turks made several attacks but were driven back. During the forenoon our troops, which have been reinforced, commenced to advance. By afternoon our troops were well up in the village. We fired 4 12in, 15 9.2in and 400 12pdr, which was rather quiet compared with yesterday.'

Above:
Landing troops at Gallipoli; *Implacable* is in the background. The battleships provided support for the troops ashore both during the landing and for many weeks thereafter. *IWM Q13220*

Right:
***Canopus*, who had arrived from the Falklands, rides at anchor in Port Iero, Mitilini, on the island of Lesvos, just south of the Dardanelles. Two monitors are also at anchor, one beyond and one astern of *Canopus*. With their shallow draught, anti-torpedo bulges and large guns, they were able to take over much of the work of the battleships.** *IWM Q13751*

Below:
***Albion* aground off Gaba Tepe being shelled by Turkish guns.** *IWM Q13808*

Above:
Attempting to tow *Albion* off. A wire has been passed to her stern; *Albion* appears to have a false bow wave painted on the waterline at her bow. A number of ships did this — it was intended to give an attacking submarine a false impression of speed. *MPL*

And in *Canopus*, who had arrived from the Falklands:
'24 Apr . . . At 5.30 arrived . . . *Doris* and *Dartmouth* closed on the shore and opened fire on the Turkish positions . . . At 8.30 our airman went up and reported no guns in the fort and no troops to be seen. During this time one of our destroyers, which was close inshore, was fired upon. She quickly replied with her guns. *Canopus* closed inshore and opened fire. 11.40 — returned to transports. 12.30 — carried out feint landing. We sent several boats ashore with troops and landed them without resistance, the operations being covered by *Canopus*, *Dartmouth* and *Doris* . . . (Later) we had news that our main army has made a successful landing.'

The routine now became one of supporting the troops ashore. On 27 April *Implacable* was amongst the ships acting in support:
'Ship bombarding Turkish trenches all day long. Landing stores, etc on "V" beach under slight shellfire. The troops made a general advance under the cover of the fire of the Fleet. As far as we know, our firing went well into the Turks, and our shrapnel did a lot of damage.'

These operations were to continue for some time. On 8 May it was *Canopus*' turn:
'May 8 . . . 12 noon weighed and proceeded up the Straits to relieve *Majestic*. At 2 o'clock our troops made an advance, the ships bombarding the Turkish positions. I had a good view of the battle ashore. It was an awful din with all the ships firing inside and outside the Straits. Where the shells burst you would think that a worm could not live, let alone a human being . . . We could see the Turks leaving their trenches and running for all their lives.'

The ships were still vulnerable to Turkish shellfire:
'May 14 . . . At intervals during the afternoon several Turkish shells fell very close, one of which went right over our ship. Several of us were watching events when this shell came whistling over us. We made a beeline for below. When we heard it coming, everyone lay flat on the deck. We expected it to burst.'

Not all the operations involved shelling enemy positions. In late May *Albion* ran aground in the vicinity of Turkish positions and had to be towed off by *Canopus*:
'May 23 . . . It turned out that *Albion* had run ashore close to the enemy's positions. *Canopus* proceeded to give assistance. We got out our 6½in wire and managed to get it onboard the *Albion* and prepared to tow her off . . . In the meantime we sent a wireless message to the Admiral onboard *Lord Nelson* who came to our assistance. At 6am the Turks opened fire on us and the *Albion* . . . Later they got more guns moved up into position and began to belt away fast and furious, many bursting over our heads. At 7am *Lord Nelson* arrived and by that time things had begun to hum a bit. *Albion*, *Lord Nelson* and *Canopus* opened fire on the Turks' positions with 12in and 6in guns. At 10.30 we got the *Albion* on the move. At 10.45 we had her off and out of range, not a moment too soon. The *Albion* was hit 40 times and had 2 killed and 9 injured. Marvellous to say, *Canopus* was not hit, though the shells fell all around.'

On 13 May *Goliath* was torpedoed and sunk by a Turkish torpedo boat as she lay at anchor. However, on 25 May a new threat arrived in the Dardanelles — the U-boat.

A group of battleships was at anchor early in the morning when a minesweeper sighted a periscope which was engaged by *Agamemnon* with 12pdrs. All the battleships started to weigh anchor and shortly thereafter the submarine fired two torpedoes at *Vengeance*, which missed. The battleships then proceeded to their bombardment positions.

At midday *Triumph* was carrying out a bombardment with her torpedo nets deployed when a periscope was sighted. The QF armament was manned. At 12.30 the periscope was sighted again off the starboard beam, then the track of a torpedo was spotted, heading for the ship. The torpedo did not explode on hitting the nets as expected, but passed straight through and exploded on hitting the ship's side. The ship took on an immediate 20° list, which then increased steadily. Destroyers quickly came alongside to start taking off men but the ship capsized and many were trapped below. In *Lord Nelson* they saw some of the survivors:
'HMS *Triumph* was torpedoed today off Gaba Tepe whilst lying at anchor. This caused a commotion. We greased off to Mudros, arriving at that well-known place in the evening. We were able to see some of the

Above:
The German submarine *U25* in Turkish waters. It was the arrival of submarines in the area which finally led to the withdrawal of the battleships. *Real Photos*

Right:
***Triumph* capsized and sank after being torpedoed by a submarine off Gaba Tepe.**
IWM Q23377

survivors which were brought in by the destroyer. A good many of them were in the rig of the day, ie flannel "dickies" and bathing drawers.'

Two days later *Majestic* was torpedoed and sunk off Cape Helles, again by a submarine. It was the arrival of U-boats which severely limited battleship operations thereafter. However, much of their work could be done by the newly-arrived monitors, which were fitted with torpedo bulges, and by similarly-fitted cruisers. The pace of battleship operations slackened considerably and they spent much time at anchor at Mudros, with only occasional forays.
'After spending a month at anchor, weighed and proceeded on one of our joy rides. We bombarded Chanak and as per usual were soon setting the village on fire. Received some small change in the way of shrapnel.'

On 20 June *Lord Nelson* carried out a notable bombardment of the town of Gallipoli from long range, using a kite balloon to spot the fall of shot, as described in a letter home:
'Since I wrote we have bombarded Gallipoli town early one morning. We went up there all night. At 3am the hands were called. We went to Action Stations at 3.30 and then fired 105 rounds from our 12in and 9.2in guns over the land at the town. We had the balloon observing for us and we did a lot of damage. The town was full of troops and the harbour full of shipping. We started on the ships as the town could not run away, and did a lot of damage. And then we started on the town with lyddite, I believe, and started five or six fires, some serious ones. I should think the troops had moved on by that time. It was still burning when we went away.'

The force of battleships was slowly reduce in numbers. Meanwhile, on land, th soldiers were unable to make progress an eventually the troops were withdrawn i November 1915. Some battleships remaine in the eastern Mediterranean but th Dardanelles campaign was over. The fina cost was five British and two Frenc battleships lost and many others damaged one or two very seriously, others onl superficially. Constantinople was neve threatened by Allied forces.

Three
1916-18

1 The Grand Fleet

Vhilst the Dardanelles campaign was being ought the Grand Fleet, most of which was based at Scapa Flow, had settled down to a outine. Midshipman Davis, then in *Neptune*, gives a good account of the outine:

. . . for recreation we went for walks on the slands surrounding the Flow. Sometimes we played golf, usually with some borrowed lubs, on the island of Flotta. The walks were great fun and not infrequently we persuaded some of the local farmers to give us tea with lots of wonderful baps, scones and cakes. These generous farm folk not infrequently refused to take any fees from Midshipmen. There were few if any games acilities at Scapa so we had to make our own recreation. We got ashore whenever we ould.

'Ships exercised underway with what were known as sub-calibre exercises. This consisted of fitting the main armament with small ifles, with ranges of up to about 5,000 yards, iring small solid shells weighing some 6 pounds. This way all the control can be exercised without properly firing the guns and their shells weighing something like 850 pounds. Then every few weeks we would steam out of the Pentland Firth to carry out a full calibre firing against a battle practice target. Firing of our 12in guns, sometimes all together as broadsides, with muzzle velocities of 2,800 feet a second, made a tremendous and quite frightening noise. My first impression of hearing a full-calibre broadside was that I thought the end of the world had come, and from then on I always used eardrum protection.

'About every 6 weeks the Commander-in-Chief took the Grand Fleet to sea for what were known as "PZ" exercises. They were, in fact, tactical exercises on a grand scale, with parts of the Fleet pitted against one another. They were intended to get ships used to working together cohesively and to get to know each other, for when action is joined a large number of signals are used and these do not always get through, so it is essential to know what the Divisional or Fleet Commander wants.'

An emergency squadron was kept at short notice and regular sweeps were carried out in the North Sea. Going to sea at frequent intervals kept the Fleet battle-ready, hardened to the rigours of the North Sea in all weathers. For many men, it continued to mean long, cold, boring hours closed up in gun turrets, casemates, magazines or as lookouts, whilst below the stokers sweated away feeding coal to the boilers.

They learned much. They learned how to sleep at their stations, huddled against the cold. They learned that a 100lb round shot, heated to white heat then put in a sand bucket, made a very effective radiator. They learned that the 6in casemates, fitted to the newer classes of battleship and battlecruiser (eg *Iron Duke* and *Tiger*) were untenable in rough weather due to water washing into them. They learned that, even when they could man the 6in guns, fumes were blown back into the casemates if firing on an ahead bearing, choking the gun crews.

At a higher level, the Fleet learned to operate as a Fleet. They practised manoeuvring battleships in close formation; they practised operating in low visibility. They

Below:
Warspite **at anchor in Scapa Flow in 1916. The 'Queen Elizabeth' class, of which** ***Warspite*** **was a member, were the most modern and effective ships in the Grand Fleet.** *IWM SP836*

Above:
The Grand Fleet at sea on manoeuvres.
P. Liddle/Cdr J. G. D. Ouvry

Right:
Shells falling close to a battle practice target. The targets were simply square latticeworks set on rafts which could be towed, from a safe distance! *IWM Q20645*

Far right:
An example of the formation of the Grand Fleet (not to scale).

practised signalling by flag and by wireless; they practised tactics. They practised navigation — the sort of navigation which allowed them to rendezvous successfully in the middle of the North Sea when they had not had a fix, perhaps, for several days. (There was no radar, no radio aids in those days.) They also learned the problems of operating by night in war, with no lights showing. Faint blue lights lit the fire control instruments; a faint glow from the compass repeat might light the face of the Officer of the Watch on the bridge, but otherwise it was pitch dark. They practised with star shell and searchlights.

Keeping people informed of what was going on was always a problem at sea because there was no broadcast system running throughout the ship. Routine messages were passed by bosun's mates passing through the ship. Calls to 'Action Stations', etc were by bugle. Voice pipes connected all the key areas of the ship, along with telephones. It was over these communication routes that those who could see what was happening, eg men in gun control positions, kept those below informed, often informally, one young sailor passing 'the buzz' to his mates below.

But for the Admiral, communication was a key element in a day when radio was rudimentary and unreliable, and limited to Morse. It was not until some time into World War 2 that Admirals could manoeuvre ships using short-range voice communications. Flags were the main method of passing orders at sea by day, with radio for those ships which were out of sight. However, if groups of ships were operating within tens of miles of each other, not infrequently single ships were positioned between the groups to act as relays, passing on messages by flag or light, rather than by using radio. Not only was radio unreliable but it could also be intercepted by the enemy. So on the bridge of the Flagship, the Admiral would pass his orders to the Flag Lieutenant. 'Flags' would translate them into the appropriate code and pass the order to the flag deck to hoist the corresponding flag signal. Up would go the flag hoist, to be acknowledged by the ships to which it was addressed, and then executed when it was hauled down on the Flagship. By such means, large formations of ships were manoeuvred.

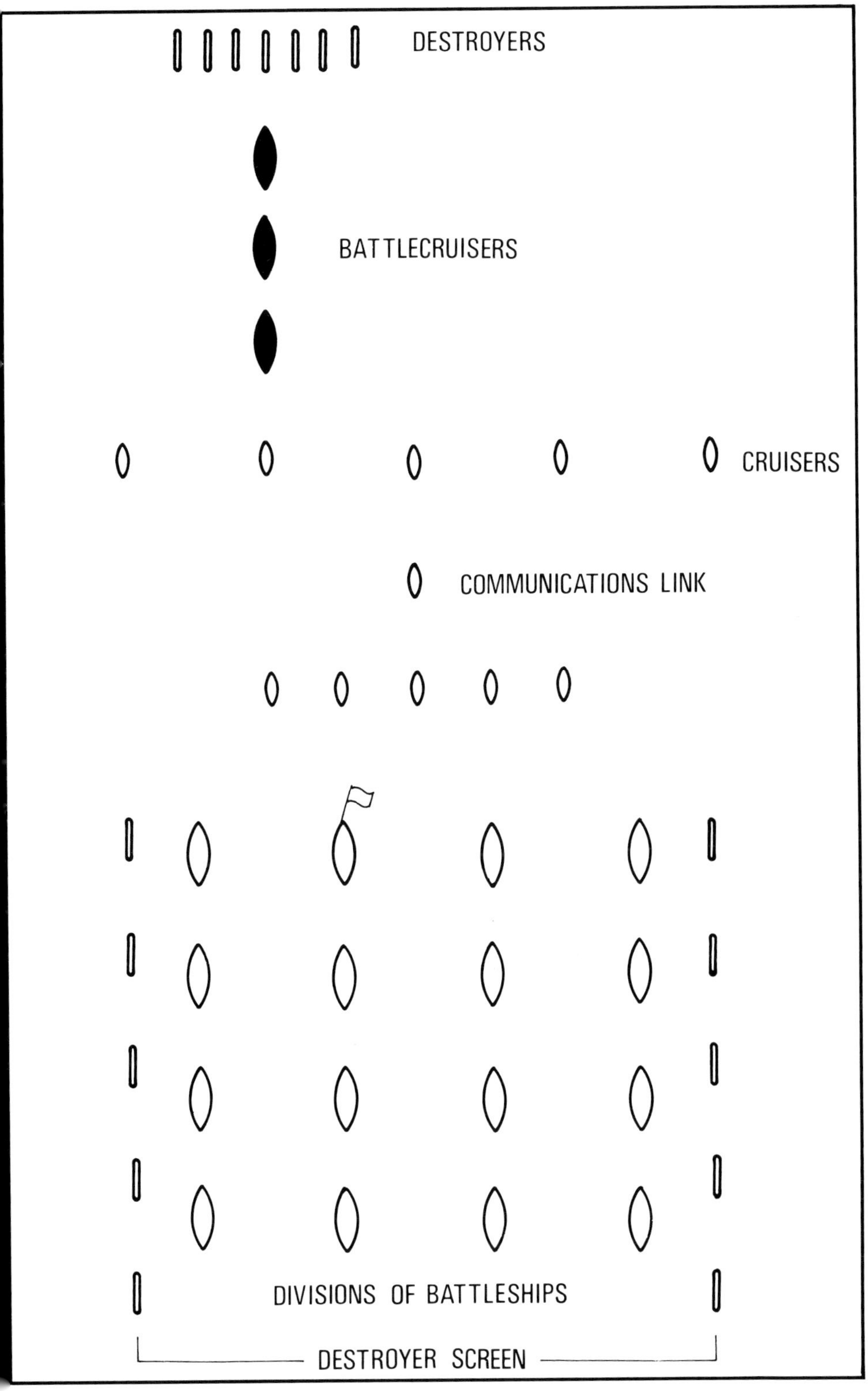

On return to harbour, one of the first requirements was usually to coal ship. This was an operation which involved the whole ship's company, officers and men alike. In fact, it was often the Midshipmen who found themselves down in the lighter shovelling coal:

'Immediately on return to harbour, the coal lighters were secured alongside and all hands coaled ship. *New Zealand* had 33 huge boilers and at 28 knots we used a lot of coal: something over 1,000 tons was an average intake. I shovelled coal into 2cwt sacks at the bottom of one of the lighter's holds. It was hard work and you had to keep your wits about you as a hoist of about a dozen sacks was no respecter of persons on its way up to the deck of *New Zealand*. The empty hook lowered for the next hoist could knock your brains in while you concentrated on the coal below you.'

The coal sacks were hoisted up on to the deck of the ship and there shifted on barrows to the top of the coal chutes. The coal was tipped down the chutes to drop into the bunkers below, where it had to be trimmed by the stokers. At the end of coaling, which would take all day in any weather, the ship and the men were filthy and both had to be washed down to get clean again. Coal dust got everywhere.

The ships lay at anchor in Scapa Flow and everything had to come to them by water. Numerous craft provided coal, ammunition, food and stores, plus transport ashore to the few facilities available. Much of the running around was done by fishing vessels, Scottish drifters, which became the work boats of Scapa Flow. In addition, ships ran their own boats for carrying officers around and going to and fro from the Flagship with written orders and reports.

Early in the war, the facilities at Scapa were very limited. However, the routine was adopted whereby a few ships at a time were sent to Invergordon or Rosyth. A Sub-Lieutenant of the time compares the two:

'One must remember that in those days when all the ships except destroyers were coal-burning, the majority of the ship's company had, perforce, to take a lot of exercise shovelling coal out of the colliers and into the bunkers. In the early days at Scapa (when possible) as many men as possible landed for a route march over the Orkney roads with a band playing good marching tunes ahead. But at Rosyth there were playing fields, a "wet" canteen where beer to each man was rationed out. Also at Rosyth the officers could get some golf . . . as well as excursions to Edinburgh, but I rather fancy that in those days, at least in the early days, the number of officers going to Edinburgh was limited.'

In time, however, improvements were made at Scapa: football pitches were constructed and a golf course built on the island of Flotta. Despite the long periods at anchor, the men kept remarkably fit. As mentioned above, coaling was arduous work and the whole existence of stokers consisted of hard physical toil. In addition, many of the tasks of ammunitioning, storing, operating guns, etc, required strength and stamina. Boxing, running, deck hockey, exercising with medicine balls, water polo in the summer, were all popular. Rowing (or 'pulling' as the Navy terms it) regattas were a source of intense rivalry between ships. The boats were not the lightweight boats of today, but the heavy 12-man cutters or 5-man whalers which required considerable strength and stamina on the part of the oarsmen to race over long distances; 5-mile races were not uncommon.

Messdeck life was very similar in World War 1 to that which another generation found in World War 2, and will be covered in more detail in Chapter Five. Life was basic: each man slept in a hammock, and the mess furniture consisted simply of a table with benches down each side. With a dozen

Top:
The Grand Fleet at anchor in Scapa Flow. The ships could not go alongisde, so all stores, ammunition, coal and men had to be transported by boat. *IWM SP407*

Above:
The Grand Fleet at sea. It is clear from this picture that the ships could be well spread out, which made the business of communications before the days of voice radio at sea all the more difficult. *P. Liddle/Cdr J. G. D. Ouvry*

Right:
The problems of fighting by night were considerable. This is *Marlborough* illuminated by searchlight. *IWM Q18592*

or more others, a man had to share all his off-duty moments, with little privacy. Early in the war it was still the practice in some ships for men to keep all their kit in a kit bag, although lockers were introduced in time for all men. Even so, the stowage space available was very limited. Despite the cramped conditions, there was a strong camaraderie amongst the men — feeling of dangers and discomforts shared.

The men were not well paid and for married men in particular the money would not go far. There were many 'firms' on board — men who would set up as barbers, tailors, cobblers, washers of clothes, even hammock-slingers. They charged for their services and were able to make a little more money. There were even firms who would 'do anything', including cleaning and polishing kit, and even hiring out particular items of clothing. Gambling was another means by which some men enhanced their pay. It was strictly illegal but games of 'crown and anchor' still went on in the evenings, with sentries posted.

Writing letters was an important pastime; an efficient mail service was run to the ships of the Fleet. In the days before radio and television, service newspapers were used to pass on world and naval news to the men. In addition, the exchange of news and 'buzzes' (rumours) between ships by boats' crews was an important route by which the lower deck exchanged information.

Cinema shows, whist drives, concert parties, the playing of musical instruments, and even dancing classes, were all off-watch leisure activities. Concert parties, in particular, could become lavish affairs with much effort put into their production. Indeed one ship, the *Gourko*, was fitted out as an entertainments ship with a stage and seating in what was effectively a floating theatre. This vessel secured alongside each battleship for a few days at a time to allow concert parties to be staged. Also, the *Borodino* became a canteen ship, providing entertainment of a more liquid variety.

The whole Fleet anchored in Scapa Flow was an impressive sight; each battleship and battlecruiser was a floating town in its own right. One officer, coming to join *Neptune* late in the war, describes the sight:

'What a sight met the eye as we steamed through the channel between Flotta and Foula. Row after row of battleships. There is the Fleet Flagship, HMS *Queen Elizabeth*, with Admiral Beatty flying his flag with the 1st Battle Squadron of *Revenge* and 4 other "R" class and 4 "Iron Duke" class battleships. Then the 2nd Battle Squadron with Adm de Roebeck in a "King George" class with four other of the class, and 4 "Orion" class battleships, *Erin* and *Agincourt* . . . Then came our own Squadron, the 4th Battle Squadron of eight ships, all armed with 12in guns; anchored a little further out in the Flow were the 4 battleships of the 5th Battle Squadron with *Barham* as Flagship. They are our latest ships and, with their 15in guns, are probably the most formidable Battle Squadron in the world. Then a number of Cruiser Squadrons, "Southampton" class, were anchored on the flanks of the battleships. It was an absolutely breathtaking sight, all anchored in this inland Flow about 15 miles long and 9 or 10 miles wide.'

Battleships and battlecruisers were nothing without their guns, and thus gunnery played a most important part in daily life. Gunnery was practised regularly: sub-calibre firings were used to practice gunnery control as full-calibre firings from the main armament placed unnecessary wear on the gun barrels. Nevertheless, full-calibre firings were also carried out from time to time. A typical day of practice aboard *London* went as follows:

'3 and 6pdr sub-calibres were used and about 20 rounds were fired from each gun in each battery, and from the left gun of each turret. We straddled the target with the second salvo and after the third went into "independent" and remained on the target. Then we went into "armoured control" and then the practice fell off considerably. The last few rounds were fired in local turret control. In the afternoon we did our full-charge firing: the Divisions steamed past, each firing at their opposite numbers' target. The Red, White and Blue watches were all much of a muchness. The main problem was that the spotting corrections were too small, and too many of our shots went over . . . There were several misfires due to the gunlayers not pressing the triggers sufficiently hard. This may also have been partially due to the amount of grease.

'In the afternoon we went back to our first formation for 6in and 12in full-calibre firing with reduced charges. Both Divisions fired simultaneously and the 12in projectiles made a noise exactly like a slow-moving railway train rumbling past. The 12in shooting was exceptionally good and straddled almost immediately and stayed on the target throughout the run.'

The gunnery system aboard *London* was not the most modern; in the more modern ships it was becoming far more scientific and gunnery control a more complex business. The Rate Officer of *Valiant* describes how the system coped with the problem of trying to hit a moving target from a platform which was itself moving, and rolling and pitching as well:

'My action station in *Valiant* was Rate Officer in the 15in Gun Control Tower, a

Above left:
Coaling from a collier alongside *Australia*. The ship's crew provided the labour on the collier as well — these were often Midshipmen.
IWM Q18753

Left:
The Commander is checking off the scores — the number of bags of coal handled — from each team. There was always an element of competition in coaling, not only between teams on the ship, but also between the ships themselves. It was the only way to get a back-breaking, arduous task done as quickly as possible. *IWM 13779*

Above:
The 6th Battle Squadron going to sea from Rosyth. A break at Rosyth offered more amenities than the bleak Scapa Flow. This photograph must have been taken quite late in the war as the ships on the left are American, with their characteristic lattice masts. Either tripod or lattice masts were used so that if part of the mast was shot away, the whole structure would not collapse.
P. Liddle/Vice-Adm E. Langley Cook

Below:
A tailor at work in *Royal Oak*. Many men made extra money by setting themselves up as 'firms' — tailors, cobblers, barbers, etc. *IWM Q17955*

Below right:
Physical training on the forecastle of a battleship. Despite spending most of the time on a ship, the men were very fit. *IWM Q17974*

cylindrical armoured tower set more or less in the centre of the Conning Tower which surrounded it. On top of the Gun Control Tower was mounted the armoured director which duplicated the Director Tower set in the foretop. The rate of change of range was an important item in the gunnery control set-up. It was up to me to try to estimate it at any time. My instrument was what was known as a "Dumaresq" after its inventor. After I had made an estimation of the inclination of the target, ie the angle which the target's course made with my line of sight, and estimated the target's speed, I set these on the Dumaresq and was rewarded by being told the rate and deflection left or right to be set on the gun sight.

'But in the modern *Valiant* there was an elaborate and valuable check provided upon the 20-year-old Sub-Lieutenant's judgement. Down below in the Transmitting Station, presided over by the Instructor Commander, was a much larger Dumaresq and the rate it was showing was connected to a pencil which drew a line on the plotting paper on which the ranges of the target were being plotted as they came in. If then my rate produced a line which coincided with change of range shown by the mean of the rangefinder's plot, then my "guesstimation" of the speed and inclination of the target must be fairly good. This whole set-up was known as the "Dreyer Table" after the name of its inventor. To enable me to chat back

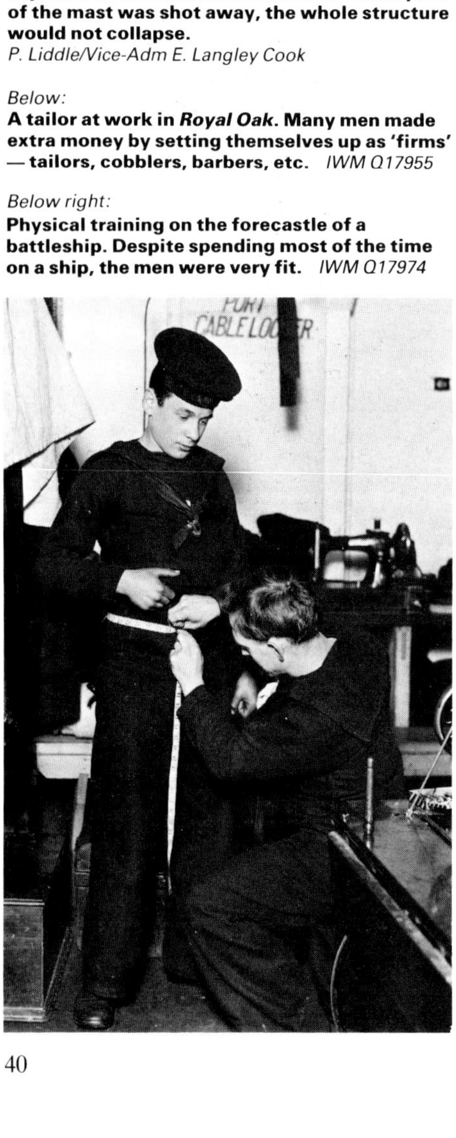

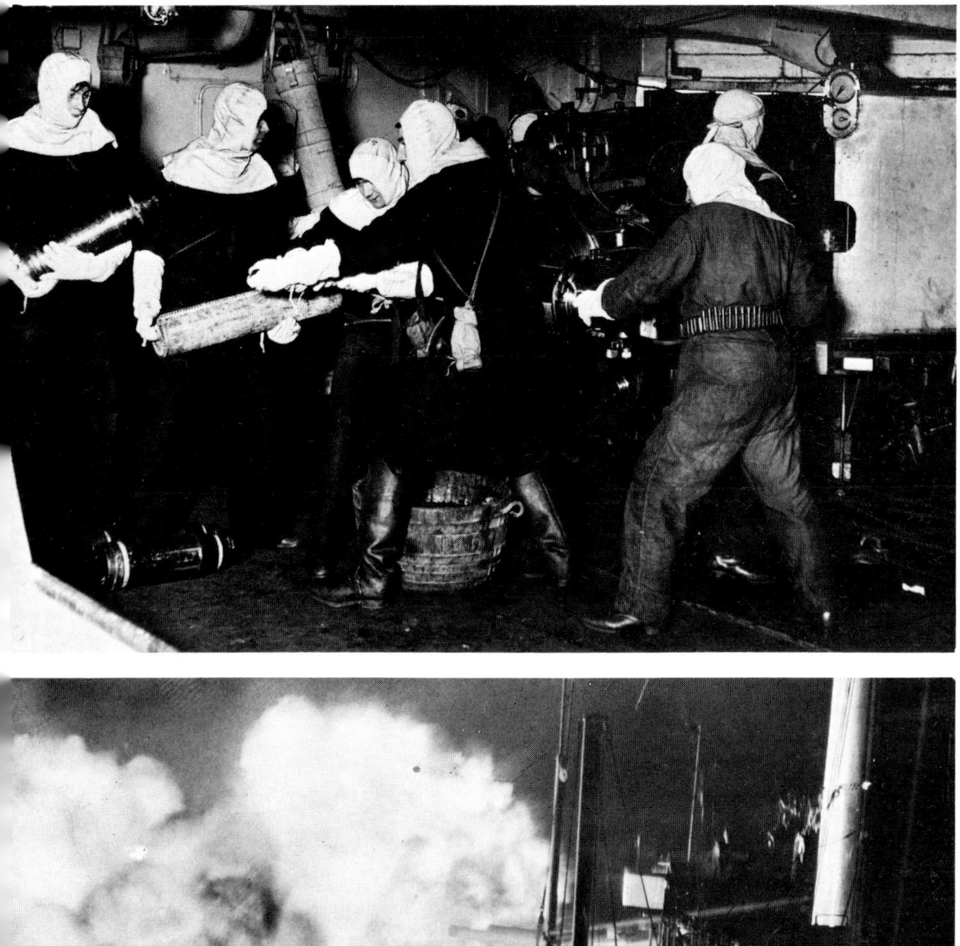

Above:
The awesome power of a battleship: *Royal Sovereign* carrying out a practice firing towards the end of the war. *MPL*

Left:
A 6in gun crew in action; in most of the secondary batteries ammunition was loaded by hand — there was no power loading.
IWM Q18005

Below left:
***Hercules* firing a 12in salvo. The gunnery system depended upon visual sighting so night fighting proved fairly difficult.** *IWM Q18033*

and forth to the Dreyer plot, I had a direct and personal telephone to the Midshipman who could make suggestions to me, guiding my guesstimations from what he deduced from his plot. I was seated immediately alongside the 15in Control Officer, so I could ask his opinion of the inclination and speed and if, from his fall of shot, it seemed that my rate was not working right, he could quickly tell me so. The rate which was decided upon at any moment was that which was transmitted to the director layers and the guns, in case the director broke down only, of course.

'Though there was a Control Officer in the foretop, the Gunnery Officer in the Gun Control Tower was regarded as the primary control and the director could be laid either from the foretop or the armoured director on top of the 15in Gun Control Tower. The 6in guns were controlled from an armoured space on either side of the main Gun Control Tower, but with no physical connection with it. Torpedo control was exercised from a Torpedo Control Tower abaft the mainmast.'

Turrets could be fired in local control, but normally the director aimed the whole main armament. The gunlayer in each turret followed the pointer on the director dial in

Above:
'Then would come the Ding, Ding of the Fire Bell. . .' *Royal Oak* firing a 15in salvo during a battle practice firing. *IWM Q18141*

Right:
***Repulse* — Gun Control System.** *Real Photos*

1 15in Director
2 4in Director
3 Armoured 15in Director
4 Triple 4in Mounting
5 Twin 15in Turrets
6 Conning Tower (Gun Control Tower within)
7 Upper Bridge and Compass Platform
8 Spotting Top

front of him, operating a brass handwheel to elevate the guns in accordance with the pointer's instructions. The trainer did a similar task to keep the turret on the correct bearing. The Nos 2 and 3 of each gun operated the gun-loading cage, the cordite hoist and the breech. The No 4 operated the rammer which thrust the shell and cordite charges into the breach. Each turret was supervised by the Officer of the Turret, sitting high up at the back. There was also an Ordnance Artificer and Leading Torpedoman (Electrician) on hand in case of technical problems.

On the order 'Load, load, load!', the cage would come up through the flash-tight doors, with much clanging, to position itself behind the breech. The shell and cartridges moved on to the loading tray and were then rammed home; the cage then descended to be refilled. Meanwhile, the gunlayer and trainer would be continuously following their pointers.

At the order 'Salvoes', the guns would be reported ready. Then would come the 'Ding, Ding . . . Ding' of the fire bell and then, with an almighty crash, the guns would fire. The whole gun would recoil some 4-5ft and, as the gun ran forward again, the breech would open and jets of water would spray on to the breech to keep down the cordite fumes. Nevertheless, and despite the fitting of ventilation, the turrets would still fill with the fumes; what with the roar of the guns at each salvo, it was exhausting work for the crews. Occasionally an individual gun might miss a salvo if it had not completed loading for some reason, but in general the firing went on a steady rhythm until there was a lull, or action ceased.

In general, the teaching of the day under estimated the ranges at which gun actions would be fought and overestimated the accuracy of gunnery. Accuracy on opening fire was important, particularly if the enemy was only in sight for fleeting moments. Such accuracy required precise rangefinding; unfortunately the British rangefinders at the time were not so good as their German counterparts, and the British were often surprised by the accuracy of German salvoes early in an action.

2 The Battle of Jutland

The great test of gunnery was the Battle of Jutland. It was a test of many other things as well — the skills of the Admirals, tactics, signalling, damage control, morale, efficiency. The course of the battle has been fully described in many books, but a flavour of the day can be gained by looking at just one squadron — the 5th Battle Squadron.

The 5th Battle Squadron (5BS) was commanded by Rear-Adm Evan-Thomas in *Barham*, with *Valiant*, *Warspite* and *Malaya* forming the rest of the Squadron. These were the latest battleships to join the Fleet, battleships of the 'Queen Elizabeth' class, armed with 8×15in and 14×6in guns, with strong armour and oil-fired boilers providing steam to turbines which gave them a speed of 24kt. 5BS was attached to the Battle Cruiser Force under Vice-Adm Sir David Beatty in *Lion*.

It was the Battle Cruiser Force which first joined action with the Germans on that fateful day, 31 May 1916. Initially it was battlecruiser against battlecruiser as the German ships attempted to draw Beatty to the south, after the initial sighting at 3.30, to where the German High Seas Fleet lay. In the early stages of the action, 5BS was somewhat astern of the battlecruisers and unable to engage; however, by 4pm they had closed the gap. It was the task of the Assistant Paymaster in *Malaya* to keep a narrative of the action:

'At 4 o'clock *Barham* opened fire at the enemy on our port bow, our BCs (battlecruisers) on starboard bow.

4.02 *Valiant* fired.

4.05 Large explosion in ship on star-

Top right:
Adm Sir David Beatty who commanded the battlecruiser force at the Battle of Jutland when a Vice-Adm. *P. Liddle/Cdr J. G. D. Ouvry*

Above right:
The 5th Battle Squadron (5BS) led by *Malaya*. The ships of the 'Queen Elizabeth' class made up this squadron, which operated with the battlecruisers. *MPL*

Right:
A further shot of battlecruisers at sea. Two 'Indefatigable' class are centre and left-of-centre, with at least two 'Lion' class beyond; then one, maybe two, 'Invincible' class on the right. The 'Invincibles' were not part of the Battlecruiser Force; they formed part of the Grand Fleet. *MPL*

Above:
The 5th Battle Squadron: with the battlecruisers, these ships bore the brunt of the fighting. Considerably modified, they were to serve with distinction in World War 2 also. *MPL*

Right:
***Barham*, with a torpedo boat destroyer close on the starboard bow.** *IWM Q17997*

Below right:
The black-plumed Battle Fleet at sea as they must have looked when closing the German Fleet. The battleships are in columns in this photograph, but by the time they engaged the German ships, Jellicoe had disposed them in a single line. *MPL*

board bow. *Warspite* fired. Destroyers on our starboard beam blowing off steam.

4.10 *Warspite* fired.

4.15 Either A or B turret, or both, fired. Range 20,000(yd). Fell short and to the left. Spotting hindered by the thick, black smoke from *Warspite*, the wind being on starboard bow, and enemy before port beam.

4.20 Ordered to engage second BC in enemy's line, but did not fire. Torpedo boat flotilla fallen back to our starboard quarter.

4.25 Either A or B fired at rear enemy ship. Short. *Valiant* hauled out to starboard. Enemy rear ship now about 5 points off bow. Range 16,600.

4.26 Splash about 500 yards short of us.

4.30 Remainder of Squadron now firing continuously. We did not fire as much.

4.32 A or B fired. Short. Range 19,000.

4.35 Enemy turned away to port. Enemy destroyers in sight on port bow far away.

4.36 Enemy on port beam.

4.40 Fired. 20,000 yards. Over and a little to the left.

4.42 *Warspite* hauled out to starboard, dropping back on our beam.

4.45 Salvo of large calibre fell about 1,000 short of *Valiant*, followed by one of smaller calibre.
4.47 Salvo fell just over *Valiant* and another just ahead of us, followed by several more over us, apparently firing common shell as of those that burst on the water; about 50% made thick, black smoke. We zigzagged slightly. Range still 20,400 . . .
4.58 Altered course 16 points and followed battlecruisers.
4.59 Salvo fell about 50 yards over us.
5.00 One shot fell a yard or two short of *Barham*.
5.05 They have got our range exactly now.
5.10 Salvo 20 yards short of us.
5.12 Straddled *Valiant* and ourselves. We are now outlined against a bright yellow horizon but they are nearly hidden in the mist.'

The rather clipped narrative disguises the fact that around 5pm 5BS had a very tense time. Shortly before 5pm, the German High Seas Fleet had been spotted coming up from the south and Beatty had ordered his ships to turn to the north in order to reverse the roles. It was now his intention to draw the German Fleet towards the British Grand Fleet. The laconic remark that at 4.58 they altered 16 points (180°) and followed the battlecruisers is, in fact, the point at which 5BS received the full brunt of the German guns for several minutes. *Barham* and *Malaya* were hit, *Malaya* twice below the waterline and once in the 6in battery, which caused a lot of damage. Events in *Barham* are described a little more dramatically (!):

'. . . Then we gave the battlecruisers a salvo or so, and the noise began to defy description. All of us worked on incessantly like automatic machines . . . scarcely conscious of what we were doing . . .

'The roar and shriek of the guns never ceased, and those who were in the proximity of the guns were so deafened by them that, despite protection provided for the ears, they were rather deaf for days afterwards. There was no doubt that our firing was very accurate and we must have caused a considerable amount of damage. After some time though, the Germans began to score some hits and the first shock which struck our ship was a sensation one is not likely to forget for many a day . . .

'We were realising by then that we were in an extremely tight corner. Terrific salvoes were being fired at us in rapid succession. So quick and fast were the shells coming that it seemed at one time that nothing short of a miracle could save us. The din of high explosives crashing on the water all around resounded in the very bowels of the ship in the most extraordinary way.'

By 5.30, 5BS had managed to extricate themselves from this predicament by opening the range. During the remainder of that day, when the two vast fleets finally came to grips, 5BS escaped further damage except for *Warspite*: her helm was jammed by a hit which caused her to circle in a wide arc, detached from the rest of the ships, at which point she received the full weight of German fire. Fortunately her rudder was soon unjammed and she escaped further damage. Of the four ships of the Squadron, only *Valiant* escaped damage that day. We will hear much more of these ships for they all played a significant part in battleship affairs in World War 2.

3 Post-Jutland

After Jutland, there were to be no further significant battleship actions. Jutland had seen the loss of three British battlecruisers; three more ships would be lost before the end of the war. After 1916 many of the older battleships were reduced to Reserve or non-combatant roles. A number remained in the Mediterranean and single ships operated in other areas of the world. The Grand Fleet stayed on at Scapa, the routine little-changed, still ready to take on the German fleet should it venture out again. But after Jutland the nature of the war at sea changed, and it was the submarine campaign against Atlantic shipping which would almost lead to a British defeat at sea. Indeed the final battleship casualty of the war fell to a U-boat when *Britannia* was torpedoed off Cape Trafalgar on 9 November 1918.

The only significant development in battleship warfare before the end of the war was the introduction of aircraft. These were designed to be used for reconnaissance and gunnery spotting, although their communication arrangements were rudimentary. Flying-off platforms were fitted to gun turrets, and launching became the responsibility of the turret officer. The aircraft was held in position by a slip attached to its tail. When the turret officer judged that the aircraft's engine was up to full power, the order would be given to release the slip and the aircraft would shoot down the short runway before launching itself over the sea. Once committed, the pilot had to return to the shore to land; the aircraft would subsequently be loaded on a lighter and brought back out to the ship to be hoisted aboard by crane.

The hazards of war were not the only dangers facing the men of the Fleet: influenza epidemics killed a significant number of men. In 1917:

Left:
***Britannia* lists heavily to port before sinking; she was torpedoed off Cape Trafalgar right at the end of the war.** *MPL*

Right:
A Sopwith 1½ Strutter being launched from one of the after turrets of HMS *Barham*.
Fleet Air Arm Museum

Below right:
***Iron Duke* at sea in heavy weather with mountainous seas washing over amidships.**
P. Liddle/F. W. H. Miller

Bottom right:
A peaceful but wet evening at Scapa Flow — this view is from *King George V*. *IWM Q19538*

'The Fleet at Scapa was spared whilst it spread on land, until some men returning from leave brought it into the Fleet. Now most of the ships in the Flow are flying the yellow flag for quarantine. I had a minor attack and felt as if I wanted to crawl away into some dark corner and die. Part of this time I was sent to work on a building being erected on the island of Hoy, near to the Grand Fleet cemetery, and it seemed that funerals were passing us most of the day.'

Despite the change of emphasis of the naval war towards the anti-submarine battle, ships of the Grand Fleet were still at work in the North Sea. One duty was escorting convoys to Scandinavia, which one Midshipman found an uncomfortable experience in *Neptune*:
'In addition, units of the Grand Fleet used to go to sea to cover convoys for Bergen. We did this on one occasion in company with the battleship *St Vincent*. I remember the trip vividly, for we shipped quite a lot of water. Battleships tend to go through the sea rather than over it, hence a lot of water comes onboard, some of which finds its way below. Climbing up the structure of the foremast to my Action Station in the foretop, situated at the top of the mast, was a rather frightening experience. If one fell off, one would probably fall into the sea and the chances of rescue in those sea conditions were negligible. It was also very cold and, in those days, we were not provided with proper Arctic clothing, so we relied upon the generous supplies of scarves and things provided by our families and voluntary workers . . . On these occasions we remained closed up at Action Stations from dawn until dusk, a pretty long stretch for a young Midshipman.'

And so the 2½ years of war after Jutland passed, for the battleships and battlecruisers at least, without major incident, although the men of the Fleet could not know that this would be the case until the day of the German surrender. They had to remain ready, trained and exercised throughout this period, putting up with the discomforts of the northern anchorages and the lack of leave, the cold and rain and sheer boredom. However, many were present at the grand finale: at the end of the war the ships of the German High Seas Fleet crossed the North Sea to surrender, and themselves to anchor in Scapa Flow.

WORLD WAR II

Four

World War 2 – The Atlantic

1 The Beginning

When war broke out on 3 September 1939, some ships were already at sea. *Renown* was one of them:

'After a week of activity, we left Portsmouth about 14.00 on the Saturday in a hurry, bound for Scapa, escorted by two destroyers. We went to Action Stations and that night the ship was darkened. My Action Station was in the 4.5in turret working the conveyor belt bringing the 4.5in shells into the turret. We were at Action Stations the following morning, Sunday, when it was broadcast that we were at war with Germany. That night the liner *Athenia* was torpedoed in the vicinity and the destroyers were dispatched to help; we belted on to Scapa, arriving there the following morning. Whilst at Scapa, each and every day was spent in all sorts of drills and sub-calibre firings.' *D. K. Bean*

One of the first requirements as war threatened had been to bring the ships up to their war complements. Some found the process remarkably simple:

'One morning, three classes were detailed off to muster in the drill shed; we were lined up three deep. The Master at Arms came along the line, put his hand between two sets of three and said "The party on my left report to the drafting office for HMS *Hood*". I was next in line which meant the next lot

Right:
Rodney **who, along with her sister ship** ***Nelson*****, were the only British battleships completed between the wars. With 9×16in guns, they were the most heavily armed ships in the Royal Navy.** *IWM A9604*

Below:
Renown**, who had been extensively modified between the wars, was at sea when war broke out. Astern is one of the 'Nelson' class.**
IWM A13012

was for HMS *Repulse*, the last lot for HMS *Renown*. We left Chatham dockyard by train and finished up in Scapa Flow. When we saw the size of *Repulse*, all were of the same mind — "how were we going to find our way round that big ship?" ' *G. J. Avery*

N. Hill, was one of the first 'Hostilities Only' ratings, and also in *Repulse*:

'A volunteer for the Navy on 2 September 1939 as an electrical engineer, I was recruited as a wireman. After initial training I joined HMS *Repulse* at Devonport in January 1940, along with two other class-mates from the torpedo training school . . . We were allocated to the Torpedo Division messdeck and became curiosities, since we were the first other ranks "hostilities only" ratings to join HMS *Repulse*. I recollect I was very disappointed at the very menial level of work with which I was involved. However, I had volunteered and I made the best of it.'

The involvement of battleships in the protection of convoys started early, and was to continue throughout much of the war. D. Smith found himself sailing to escort the first convoy on the second day of the war:

'Sept 4th at Weymouth. The *Ramillies* and two destroyers, *Exmouth* and *Escapade*, were ordered to sea just after 21.00. After passing the breakwater, *Ramillies* signalled to the destroyers to take up stations for the night, one destroyer on each bow at 45°, 1,200 yards, thus giving *Ramillies* protection against any U-boats. Night lookouts, search-light crews and a section of the gun crews took over their duties. Our course was westward until, at midnight, we altered to the northwest at 10 knots. Each of us kept wondering where we were bound.

'Dawn came and the group was still steering northwest at 10 knots; the weather was unsettled with patches of fog. About 20.00 the lookouts sighted smoke in between the patches of fog . . . and as the fog cleared we could see about 16 ships. Of course, we knew now that this was to be our convoy. At present the convoy was being escorted by 9 destroyers; *Ramillies* and her destroyers joined, so our convoy and escort were all formed-up steaming southward. The course was a zigzag from now onward. The ships were in four columns with the *Ramillies* right astern in a position where all the ships could be seen. Destroyers kept dashing around the convoy with their Asdics pinging away . . .'

The convoy, which was the first troop convoy, subsequently reached Gibraltar safely, with many false alarms for sub-marines on the way and one possible submarine kill. For *Ramillies*, there was little to do.

The threat to the convoys, which neces-sitated the use of battleships in the escort, came from German surface raiders. At the beginning of the war two such ships were at sea: *Deutschland* was operating in the Atlantic and was the ship which most concerned *Ramillies*. In addition, the *Graf Spee* was on the loose in the South Atlantic.

The ships of the 'Deutschland' class were unique because they were designed with commerce raiding in mind. In Germany they were initially known as *Panzerschiffe* or

Above:
Part of the first convoy of World War 2 which was escorted by *Ramillies*, from which ship this photograph was taken. *D. Smith*

Left:
The sleek lines of *Graf Spee*. Armed with 6×11in guns and with a maximum speed of 28kt, the three ships of this class were ideal for commerce raiding. Some speculation has arisen over whether or not *Graf Spee* had radar. This photograph shows it clearly — the rectangular array on the director high on the foremast. *Real Photos*

'armoured ships', but they were popularly known as Pocket Battleships. The three ships, *Deutschland* (later renamed *Lutzow*), *Graf Spee* and *Scheer*, were armed with 6×28cm (11in) guns as main armament in triple turrets: one forward one aft. They were propelled by eight MAN Diesels driving two shafts, which gave them a good range and a maximum speed of 28kt. For their size they had good turret, deck and belt armour. They could outgun any cruiser and outrun any battleship. The problem of dealing with them was not dissimilar to the problem of Von Spee's Squadron off the Falklands in the previous war — only battlecruisers could match them for speed and outgun them. However, in 1939 the Royal Navy only had three battlecruisers — *Hood*, *Renown* and *Repulse*, and for the first few months of the war *Graf Spee* was to prove a very elusive foe.

2 The Battle of the River Plate

The events surrounding the Battle of the River Plate interest us in this book mainly from a tactical point of view. The conduct of operations by the *Graf Spee* brings out many of the tactical considerations which affected battleship warfare, in particular their use by the Germans for commerce raiding. It is also interesting to see some of the tactical problems facing the British cruisers.

Commerce raiding by a single ship in the South Atlantic was unlikely to be decisive, but it would force the British to adopt convoying over a large area, tying up a disproportionate number of escorts which were, as ever, in short supply. The mere presence of a single ship would be enough to disrupt merchant traffic. The directive to the Captain of *Graf Spee*, Kapitän zur See Langsdorf, was 'the disruption and destruction of enemy shipping by all means possible'; he was ordered to commence operations on 25 September. The South Atlantic is a vast area of ocean in which to hunt, so Langsdorf made considerable use of his aircraft, an Arado Ar196, for search purposes. (It could also act as a target to train anti-aircraft gun crews.) In addition *Graf Spee* was fitted with an early form of German radar, DT-Gerät, which had a maximum range of 18½ miles.

Langsdorf was basically searching for the single, unescorted merchant ships. He wished to avoid action with any naval forces

Left:
Adm Sir Henry Harwood when C-in-C Mediterranean. As a Commodore, he commanded the British ships at the Battle of the River Plate. *IWM A13964*

Below:
***Exeter*, the only 8in cruiser in Force 'G' which took on the *Graf Spee*. *Exeter* was badly damaged in the battle and had to retire.**
Real Photos

espite his superiority in guns over most hips, as even one lucky hit on his ship could ave serious implications. *Graf Spee* was perating a long way from any friendly bases nd repair facilities. Langsdorf found his rst victim off the coast of Brazil at the end f September; he then moved to the Cape of Good Hope-Freetown route where he found further four ships. To throw the hunters off he scent, he then rounded the Cape and ank one small tanker off the southeast coast f Africa in mid-November. Doubling back n his tracks, he found two more targets on he Cape-Freetown route before deciding to ead for the area off the mouth of the River Plate. Langsdorf was already thinking of eturning to Germany, aiming to arrive ometime in January. His ship would have een at sea for six months continuously by hat time, a very long spell by any standards, nd would be in need of a comprehensive eriod of maintenance. Also, he knew that he efforts of the hunters would be ncreasing all the time. Indeed, he had lready been saved from a brush with a British cruiser only by the fact that his ircraft spotted the British ship in time for *Graf Spee* to take avoiding action.

His method in dealing with merchant ships ad been to approach end-on to avoid being dentified until as late as possible in the ncounter. This reduced the time any ship ad in which to radio a report. He always allowed the crew to leave their ship before he sank it and was meticulous in looking after his prisoners. Indeed he kept, firstly, *Newton Beach* and, later, *Huntsman*, as prizes in which to accommodate the prisoners. However, when the time came to start preparations for returning to Germany, all the prisoners were transferred to the tanker *Altmark* and the remaining prize, *Huntsman*, was sunk.

Altmark was the key to the *Graf Spee's* operations. She was both a tanker and a stores ship, and her use for replenishment at sea preceded the massive use of such methods, particularly in the Pacific, later in the war. Fuelling was carried out using a hose from the tanker's stern to *Graf Spee's* bows, but stores could only be transferred by boat; a major top-up on 16 October took all day by such means. *Graf Spee* fuelled from *Altmark* for the last time on 6 December 1939 and headed west.

She found one more victim en-route to the River Plate and arrived in the shipping lanes off the mouth of the river on 12 December. Meanwhile Cdre Henry Harwood had assembled Group G, consisting of the 8in cruiser *Exeter* and the two 6in cruisers *Ajax* and *Achilles*, off the Plate in anticipation of such a move by the *Graf Spee*. Harwood had given considerable thought to the problems of how to take on the German ship. His cruisers were out-ranged by the *Graf Spee's* 11in guns; however, with three ships he could split the fire of the enemy and, with his superior speed (*Graf Spee* was limited to about 24kt by her dirty bottom), he had control of the range. He rehearsed his tactics and decided that *Exeter* would operate alone on one side of the *Graf Spee*, and *Ajax* and *Achilles* would act as a pair in close company on the other side.

Early on the morning of 13 December, the British ships had just fallen out from Dawn Action Stations and gone to Day Defence Stations, in which only a proportion of the armament was manned, when smoke was sighted to the northwest. *Exeter* was dispatched to investigate. Meanwhile, in *Graf Spee*, *Exeter* had been recognised, but *Ajax* and *Achilles* were thought to be destroyers. The interesting question at this point was why did Langsdorf choose to fight when his previous tactics had been to avoid action at all costs? It is said that he appreciated that, once spotted, he could be shadowed by the cruisers whilst they called in heavier ships. It has also been suggested that he chose to fight because he believed the British ships were protecting a convoy.

The British ships implemented Harwood's tactics, *Exeter* remaining southeast of *Graf Spee*, whilst *Ajax* and *Achilles* worked round to the northeast. In this situation, Langsdorf could only effectively engage one British ship at a time. It has been suggested that he split the fire of his 11in turrets but this is not so; indeed for quite long periods of the battle, with the British ships on his quarter, the forward turret could not bear on any of the cruisers. Neither did he keep changing targets: changing targets is no quick and easy evolution. It requires the range of the new target to be found and it will certainly require time before effective fire is possible against the new target. Therefore Langsdorf chose to take on *Exeter* with his 11in guns, hoping to keep the 6in cruisers at bay with his 5.9in guns which had almost equal the range.

Langsdorf's initial tactics were very successful. Again, like their predecessors in World War 1, the British found the German gunnery remarkably accurate; *Exeter* was soon hit heavily and, by 30 minutes into the action, had only her after-turret operating and was seriously on fire. However, *Ajax* and *Achilles* were hammering away with 6in salvoes and scoring hits, which although not penetrating *Graf Spee's* armour, caused significant damage to the superstructure and amongst her secondary armament. Even before *Exeter* was out of action, Langsdorf had turned away in the face of the 6in gun threat, and also the threat of torpedoes from the cruisers.

Despite the fact that he knocked *Exeter* out of the battle, Langsdorf faced considerable problems. The damage which had occurred to his ship (although on the face of it superficial inasmuch as his main armament was intact and his ship was still being propelled by her Diesels) was enough for

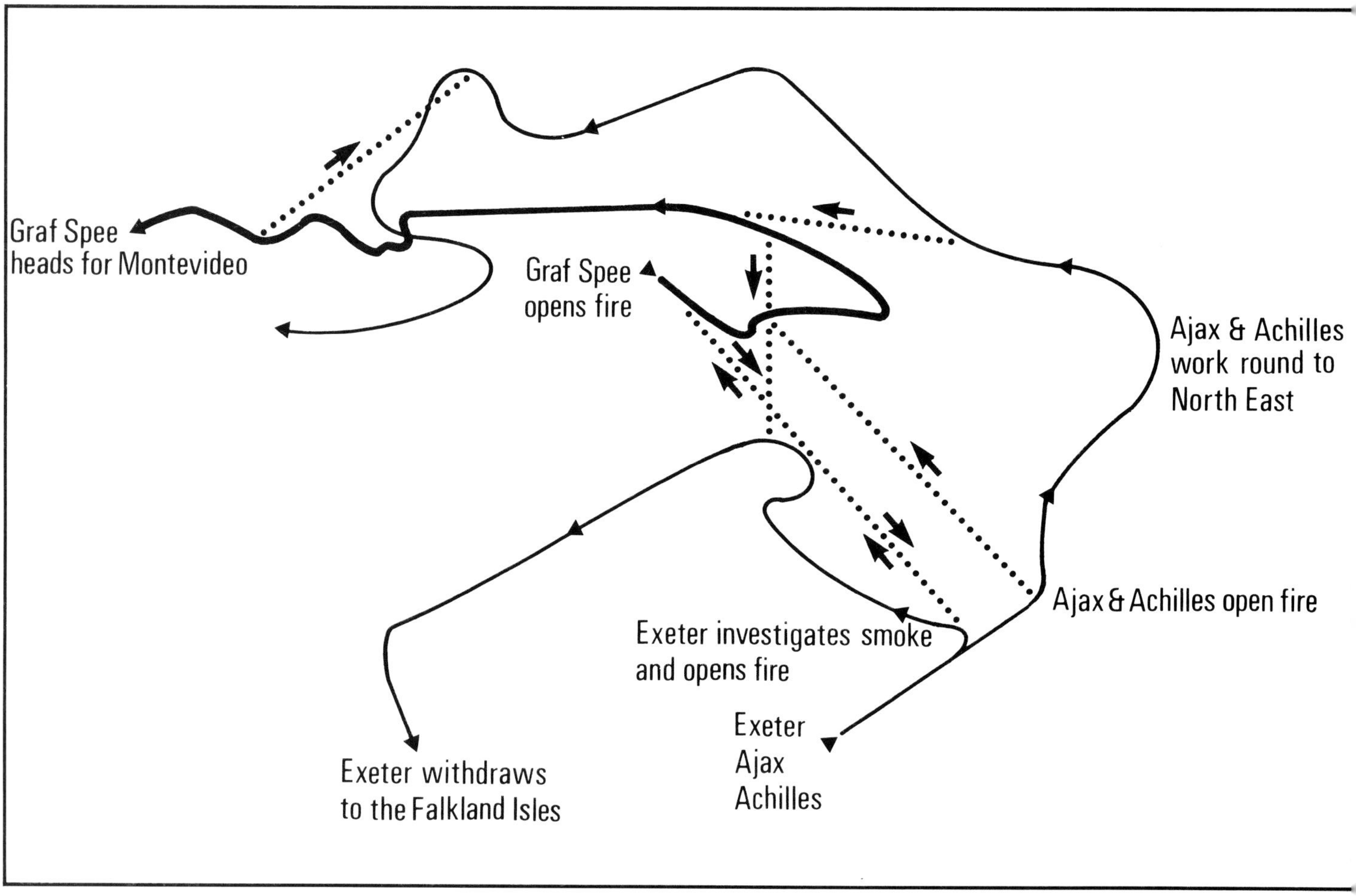

Above:
Battle of the River Plate.

Right:
***Ajax*, a 'Leander' class cruiser, was Cdre Harwood's flagship at the Battle of the River Plate.** *Real Photos*

Below right:
***Achilles* served with the Royal New Zealand Navy until 1943.** *Real Photos*

him to believe that repairs would be necessary before a return to Germany could be possible. A number of hits on the hull, particularly one near her bows, made his ship unseaworthy for the North Atlantic in winter. Also, all his galleys, except the small Admiral's galley, were out of action; some of his food stores were flooded, his main engine luboil separators were damaged and some of his 5.9in ammunition hoists were out of action. An additional factor was that he only had 40% of 11in ammunition remaining and 50% of 5.9in, therefore he chose to seek the shelter of a port to effect repairs. There was no German base anywhere in the area so he headed for the neutral port of Montevideo, shadowed all the time by *Ajax* and *Achilles*, who were in fact facing their own problems. the Captains of both ships believed they were low on ammunition, although a subsequent muster showed that they had overestimated expenditure. Also, both the after-turrets of *Ajax* were out of action after an 11in hit.

Above:
***Graf Spee* at anchor in Montevideo. Very little of the damage can be seen in this shot; note the false bow wave painted on to give submarines an incorrect impression of speed.** *IWM HU610*

Right:
***Graf Spee* showing some of the damage amidships; the Arado Ar 196 spotter plane is burnt out. The guns are 150mm (5.9in).**
IWM HU206

Subsequent events after the arrival of *Graf Spee* in Montevideo are really outside the scope of this book, except to say that the lack of adequate repair facilities were an important element in the final outcome, for the Uruguay Government refused to allow Langsdorf enough time in harbour to effect repairs. In addition, British propaganda and mis-information had convinced the Germans that they would be up against a superior force if they came out (the idea that the British had managed to concentrate reinforcements off the Plate having taken root in German minds). In fact, Harwood had only been joined by the 8in cruiser *Cumberland*. *Ark Royal* and *Renown* were on their way, but had not arrived by the time Langsdorf sailed on the evening of the 16th to scuttle his ship within sight of the British cruisers. And so the first of a series of campaigns against the German raiders came to an end.

3 Home Fleet

1939 and early 1940 were known on land and in the air as the 'phoney war', but at sea there was nothing phoney about it. During this phase most of the battleships were in the Atlantic. The Home Fleet consisted of *Barham*, *Malaya*, *Royal Oak*, *Renown*, *Repulse*, *Hood*, *Nelson* and *Rodney*, plus *Revenge*, *Royal Sovereign*, *Resolution* and *Ramillies*, the latter four bearing the brunt of convoy escorting in the early days. *Queen Elizabeth* was undergoing reconstruction and *Valiant* was in refit; *Warspite* was in the Mediterranean and the 'King George V' class were still building.

As in the previous war, Scapa Flow became the Home Fleet base and the ships settled down to the routine of wartime operations. *Repulse*, for example, found herself involved in the escorting of convoys: '. . . Oil ship, take on supplies, put to sea to take over a convoy brought out of Greenock

Left:
***Royal Oak* which was torpedoed on the night of 13/14 October 1939 in Scapa Flow by *U47* (commanded by Lt Gunther Prien).** *Real Photos*

Below left:
A 'King George V' class battleship at sea in moderate weather. Even in these conditions the ships could be rather wet forward.
R. Duquemin

Right:
At sea in rough weather — a view from *Revenge*. *E. R. Moyle*

Below right:
A High Angle Director, designed to control the guns against aircraft targets; this was a very exposed position high in the ship and the men needed to wrap up against the cold. *IWM A8139*

by destroyers. The destroyers would then return to Greenock and we would then sail for Canada. A day's steaming from there, we would pick up a convoy going to Scotland and turn over our convoy to the destroyers going back to Canada, which meant that our only harbour would be Scapa Flow, where we would oil and supply ship again ready for the next convoy. If you were lucky enough to be off watch, you were allowed ashore for three hours and rationed to three pints of beer for each rating . . . We left harbour one evening at 11pm with destroyer escort for a patrol. 2am the following morning HMS *Royal Oak* was sunk in Scapa Flow . . . Up to that time we hadn't fired a shot in anger, but it didn't alter the fact that all the time we were at sea we were closed up at Action Stations, using your gas mask for a pillow. We never saw our hammocks until we got back in harbour.' *G. J. Avery*

Covering the western end of the transatlantic convoy route was *Royal Sovereign*: 'Our convoy duties took us from Halifax to a rendezvous 300 miles off the coast of Ireland, where we handed the convoy over to other HM ships and took over the return convoy for Halifax. The outbound convoy to England was usually ships with war material and food, ships of all shapes and sizes. Some convoys were fast, and some were slower. If we had troop carriers, it would be a fast convoy. The inward-bound convoy, which we picked up at the rendezvous, consisted of passenger liners with RAF personnel for training in Canada and evacuees on their way to Canada or the USA. From the time of leaving Halifax until our return was about 14 to 17 days, depending on whether it was a fast or slow convoy.

'The pattern of life onboard at this time was one of continuous battles to keep the messdecks and ourselves dry. The North Atlantic in winter can be very rough to say the least, and although the ship was a 30,000-ton battleship, the seas still did a bit of damage. The starboard side screen door and a 3in iron support to the boat deck were both bent, also the breakwater on the fo'c'sle sprung a few rivets and water was coming into the messdeck through the deckhead. Sea water was also finding its way down some of the air vents. The cable locker was also flooded and, although the portholes were closed with the deadlights down (this was the routine at sea), the water still found a few places to come through.' *T. W. Pope*

In *Royal Sovereign* they did not spend so much time at Action Stations as in *Repulse*: 'Our routine whilst at sea was to work part of ship in the forenoon, with part of the watch closed up at Ocean Cruising Stations. This consisted of the HA (High Angle) Director and two 4in guns and a pom-pom manned. Afternoons were spent catching up on our sleep, washing, etc. At dusk we went to Action Stations and reverted to Defence Stations. This would mean that half the ship's armament would be closed up during the night. Dawn would see us once more at Action Stations and then back to Ocean Cruising Stations. My station for cruising was the forward HA Director as layer, and my Action Station and Defence Station was gunlayer on the starboard 6in gun. Although we were on convoy duties from January until late April, we never came under attack from

the enemy. All our convoys got through safely. This was without any escorting destroyers.'

(At this stage of the war, convoy destroyer escorts did not have the range to take a convoy all the way across the Atlantic.)

Royal Oak was torpedoed by a U-boat in Scapa Flow on 14 October 1939. For a time the Home Fleet retired to other anchorages on the coast of Scotland whilst the defences of the Flow were improved. When they returned, life for ships in harbour at Scapa could be boring, although in the early months of the war the amount of time spent in harbour was very limited. Accounts of the runs ashore they were able to have are remarkably similar:
'Scapa is quite barren. My memories of a run ashore at Scapa on the island of Flotta consisted of getting into the drifter, racing to the landing stage with the rest of the fleet, there climbing over umpteen other drifters and racing up to the queue for the canteen building.' *G. A. Langridge*

'Leave for months on end was neglible. At Scapa Flow — canteen leave — only every 4th day from 13.00-18.30 — canteen only. It was either egg and chips in the NAAFI and maybe a film show, or drinking pints in the canteen. Drifters used to collect the libertymen at 18.30 from the Scapa Pier, Flotta, many roaring drunk.' *F. Troughton*

'. . . found the rickety jetty packed with what seemed to be every rating at Scapa, all in various stages of inebriation, picket boats coming alongside from all the ships at anchor. Being wintertime, it was dark and the jetty was illuminated by a single blue light. It was really almost unbelievable, bodies seemed to be jumping or falling or being thrown into picket boats — amazing that no-one ever seemed to be injured.' *N. Hill*

'Shore leave for ratings usually expired during the first dog watch, when all the drifters and ships' boats would come alongside the jetty to convey the libertymen back to their respective ships. This was often the signal for certain types to strongly express their opinion on which was the best or worst ship in the Fleet; the length of the heated discussion often depending on how long one had spent in the Fleet Canteen. It

was also quite normal to see the returning drifter loaded with libertymen circling the ship until the Officer of the Watch had decided that the men had sufficiently quietened down to be allowed onboard, whereupon most of them went quietly to their messdecks.' *G. Whyte*

The image of the runs ashore at Scapa may not accord with the reader's vision of life in the Royal Navy during war, but it was the reality of men unwinding after days at sea, suffering from cold, boredom and lack of sleep; or periods in harbour when there were very few other distractions. Or perhaps sailors are allowed to get drunk from time to time . . .?

This is an old tale, but it probably happened when all the Padres gathered on the ship at Scapa. After about the fifth or sixth Padre had been piped aboard, the PO of the Watch muttered: 'We should be bloody well 'arping them aboard, not piping them'.

The threat to the convoys was ever-present. In November the *Scharnhorst* and *Gneisenau* made a brief sortie into the North Atlantic, sinking the Armed Merchant Cruiser *Rawalpindi* before returning to Germany. This was the first operational sortie by these ships; we will hear more of them.

In early April 1940 the Germans invaded Norway. Many units of the Germany Navy were at sea in support of the Norwegian operations, including *Scharnhorst* and *Gneisenau*. In response to intelligence of considerable activity by the German Navy in the Heligoland Bight area, the Home Fleet also put to sea to counter what was initially thought to be an attempt to break out into the Atlantic by a number of German ships. On 9 April *Renown* was patrolling off Vestfiord in the vicinity of the Lofoten Islands when she encountered *Scharnhorst* and *Gneisenau*. *Renown* mounted 6×15in guns against the Germans' 12×11in, but the latter had the advantage of speed. In addition the older *Renown*, a battlecruiser, was somewhat weaker in armoured protection. The weather was atrocious, with a northwesterly gale blowing, very rough seas and intermittent snow squalls. The German ships turned away from *Renown* which promptly gave chase with little hope of catching them. However, in the time it took the Germans to open beyond the range of *Renown's* guns, the latter obtained three hits on *Gneisenau* causing considerable damage, including putting the main Fire Control Director out of action. D. K. Bean tells of his recollections:
'During this time the weather was really bad: gales and snowstorms and, at one time, it was piped "All hands clear off the upper deck", as the ship was going to alter course. All the time at sea the ship was in a certain degree of readiness, depending upon the circumstances. During the evening of 8th April, the ship went to full Action Stations,

Left:
***Nelson* in Scapa Flow. The ubiquitous drifters provided transport for men and stores.**
IWM A108

Below:
The scene is familiar — just the ships have changed. British battleships at anchor in Scapa Flow; *Nelson* is nearest the camera and *Barham* beyond. *IWM A112*

then reverted to First Degree of Readiness, which meant one could relax at your Action Station. My Action Station was in P4-4.5in turret and one could slip out of the turret into the flat outside the Admiral's quarters. I tried to get my head down just outside the turret, but the ship was rolling and pitching so badly it was impossible.

'At about 04.00, Actions Stations was piped and in seconds everyone was closed up; the *Renown* opened up with her 15in. This went on, on and off, until about 06.00. *Renown* was firing on the starboard side, so the 4.5in turrets on the port side could not fire, and I think those on the starboard side did not as the range was too far. During the engagement, *Renown* increased speed and the escorting destroyers were left far astern.

'The sea was really high and the ship was crashing, shuddering and banging, and it was impossible to tell whether it was shell hits or the sea. *Renown* had fired several rounds before the *Scharnhorst* and *Gneisenau* replied. *Renown* was hit twice. One shell went right through the mast just aft of the bridge; the other shell went in on the waterline at the stern and flooded it. It did not explode and wandered round before going out the other side.

'I believe *Renown* hit *Gneisenau* three times. During the action the German ships were running and the aft turret was firing on a fine starboard bearing, just above the Admiral's dining cabin. The blast completely blew out several of the ports.'

Later in the campaign for Norway, the German pair were to sink the aircraft carrier *Glorious* and her two escorting destroyers.

Earlier, *Warspite* had been involved in a battle of a different nature. The Germans had gathered a strong force of destroyers in the approaches to Narvik and on 10 April a

Above:
***Renown* engaged *Scharnhorst* and *Gneisenau* briefly off the Norwegian coast, obtaining three hits on *Gneisenau*. *Renown* did not have the speed to keep within range of the German ships.** *Real Photos*

Far right, top:
***Scharnhorst* and *Gneisenau* seen off the Norwegian coast. These two ships frequently operated as a pair in the early years of the war. With their 9×11in main armament they were no match for any British battleship, but they had the advantage of being able to do over 30kt.** *IWM HU2224*

Right:
***Gneisenau* can be distinguished from *Scharnhorst* by the position of the mainmast — immediately aft of the funnel; *Scharnhorst's* was further aft. *Gneisenau* saw no service after February 1942 when she was damaged by mines during the 'Channel Dash', and subsequently bombed at Kiel.** *Real Photos*

flotilla of British destroyers entered the Vestfiord. In a spirited action which became known as the First Battle of Narvik, ships were lost on both sides. Two days later, *Warspite* took part in the Second Battle of Narvik:

'It was the evening of the 12th April 1940. We were told we were going into Narvik with 9 destroyers the next day; us boys were told by our instructor to bath and put on clean gear and overalls. During the dog watches we slowed down and Adm Whitworth and his staff came onboard. Around about dinner time, Action Stations sounded off. My Action Station was in X turret cabinet, working the range and deflection clock for the turret officer. X turret became a director turret if both directors were damaged. Running along the upper deck to Action Stations we saw our escort of destroyers, three in line ahead leading us, and three on either side, so we felt well looked after.

'When the action started, we in X turret felt left out. We could hear and feel A and B turrets having a real battle, and in X turret all we could hear was our turret trainer singing out "X turret won't bear". We had a running commentary of what was going on and didn't feel very comfortable when we

Above:
***Warspite* in action off Narvik firing at the German destroyer *Erich Kellner*. *Eskimo* is on the left of the picture.** *IWM A38*

Below:
The unusual design of *Rodney* had been dictated by the limitations of the Washington Treaty, limiting displacement to a maximum of 35,000 tons. She was armed with three triple 16in turrets, all forward. *IWM A9605*

were told that the German destroyers had fired torpedoes at us. They missed and we heard them explode on the sides of the fiord.

'After the battle we anchored off Narvik and the *Eskimo*, with her bows blown off, came alongside with the *Cossack* and transferred the dead and wounded onboard. The boys' messdeck was cleared of tables and stools, and our hammocks were used on the deck for beds which was now being used as a sick bay. We steamed out of Narvik that evening and buried the dead at sea. We did get some shells away from X turret because we went back into Narvik a week later and bombarded, and I'm sure Capt Crutchly turned broadside on to give X and Y turrets a chance.' *R. Emmington*

Warspite and her destroyers sank or severely damaged eight German destroyers. In addition, *Warspite's* Swordfish aircraft which had been launched to search the side fiord ahead of the battleship, attacked and sank a surfaced U-boat by dropping a depth charge which went down the submarine's conning tower hatch!

Rodney was also at sea at this time, and on 9 April was the subject of several air attacks. A Royal Marine recollects one successful attack:

'We happened to be singled out by a lone Stuka carrying a 1,000lb bomb. During those days ships could not fire fore and aft, cutting out on a fore and aft bearing. Apparently this lone pilot must have known this: he came in at about 200 feet and a running commentary of his action was being relayed from the Spotting Officer on the bridge. Everybody froze when he suddenly announced "He has released the bomb — NOW!" We all waited for the big bang but luck was on our side. The bomb went through 4 decks (I think), just behind a 4.7in platform, making a hole some 3-4 feet in diameter. Having been built prewar, *Rodney* and *Nelson* had a 12in armour plate running the length of the engine room deck, to where this bomb broke in half, firmly embedding itself in the armour plate. On inspection, the first thing the Engineer Officer noticed was that it was made in Czechoslovakia.' *W. Handley*

After the Nowegian campaign which ended in the withdrawal of British forces the Home Fleet returned to the regular tasks of covering the North Atlantic convoys, but there were no further major actions for some months.

Five

Battleship Life

1 Battleship Layout

Let us now take a more detailed look at a battleship, at the ship herself and what she was like. R. W. Scott, who was a communications rating, joined *Warspite* after her reconstruction which took place between 1934-37, and came to know her well during the war:

'The general appearance, superstructure and armament had been modified and added to; the four 15in gun turrets remained. Three were manned by the gunnery branch and one quarterdeck turret by the Royal Marines. (The Royal Marines manned a main turret in most ships.) The 6in gun batteries had been removed from the waists and the area was free for stowage and sleeping purposes. Twin sets of Bofors had been fitted on the superstructure adjacent to the funnel on each side, and 4.7in AA guns had been mounted in the same area.

'Two hangars had been erected in the area immediately astern of the funnel where the Walrus amphibious-type aircraft were housed. A small contingent of RNAS [Royal Naval Air Service] and RAF personnel serviced and flew the aircraft. Launching was done by catapult and recovery by jib crane. Fleet spotting was the main function with some reconnaissance work.

'The galley was situated on the main deck forward, with access through hatches and steel ladders to the main messdecks and to the senior ratings' messes. The galley was pretty well equipped for the period, with oil-fired cookers, steamers, fryers and all the clutter which is essential to feed some 2,000 men. The ship's bakery was further aft, access being from the waists, where vast amounts of bread and rolls, puddings and cakes were produced . . . The officers' galley and drying room, together with the sail-maker's store, etc, were in the same area leading on to the waist deck.

'The forepeak area was fairly standard — the cable locker, 3 cells, a paint store and a small isolation mess. The heads were in this area, the junior ratings' on the port side with half steel doors, the senior ratings' on the starboard side, with full steel doors. In the next compartment aft the Petty Officers' mess and locker area were situated.

'The mess occupied the full length of the compartment. The long table and leather-covered seating were down the side adjacent to the portholes, with the pantry and serving area on the other. The portholes were a mixed blessing as they were invaluable aids to ventilation in hot climates and smooth seas, but could leak in rough weather. The whole flat was prone to flooding through fan shafting and air intakes.

'Further aft were other senior rates' messes on either side of the main passage. Leading from this flat, below were the main junior ratings' messes. The stokers, writers, sick berth attendants, etc, were on the port side. The boys, seamen, communications and torpedomen were to starboard. On its own, in a separate compartment further aft and next door to the Midshipman's flat, were the Royal Marine "barracks". Steel double-doored lockers were sited in vacant areas as were hammock nettings. Bathrooms were one deck down from the messdecks, one to each flat. In areas where water was at a premium and when at sea, bathrooms were only opened twice a day for an hour in the mornings and for the same time in the evenings. Laundry was usually done in a bucket with water from the galley. The

Below:
***Warspite* is probably the most famous of all the British battleships. She was extensively modified between the wars and served in all theatres of the war except the Pacific. She is probably best known for her time with the Mediterranean Fleet.** *IWM A9701*

Right:
The ship's bakery — in this case, all-electric. Some of the older ships still relied on coal for some of their cooking. *IWM A2896*

Below right:
The Royal Marines' mess deck in *Duke of York*. The mess is ready for Captain's Rounds, with all the mess gear laid out for inspection and the whole place gleaming. *IWM A20530*

waists were a popular venue for "dhobeying" (clothes washing) sessions.

'The mess tables were long deal structures set at right angles from the bulkheads, hooked on to an angled bracket and held in place by steel supports secured to the deckhead. These supports were usually burnished to a high degree by the duty "cooks of the mess". A leading hand was in charge of each mess; he detailed off two men per day to clean the mess, scrub out, wash up after each meal and draw and issue food supplied from the galley, draw bread from the bakery and tea, sugar and tinned milk from the dry stores. Rum was drawn usually by a leading hand or senior member of the mess.'

Passage between messdecks was difficult because of the watertight doors and splash barriers, about 3ft high, which were fitted at the bottom of each door. Barked shins and skinned legs were often the result of hurried exit from mess decks at such time as Action Stations.

Leading from the messdecks on the port side were junior officers' cabins and, amidships, various offices — writers, engineers, stores office, and access to the boiler rooms and engine rooms via airlocks and watertight doors. The main WT receiving room and the central communications office were in the after part of the area.

The after part of the ship was mainly devoted to the Midshipman's mess area, Warrant Officers' mess, and more cabins. The wardroom, Admiral's quarters and Captain's day cabin were one deck above, under the quarterdeck. Boat decks led on from the waists where cutters, whalers, etc, were slung on davits controlled by electric winches. The Admiral's barge and the power boats were slung inboard by jib cranes and housed on superstructure areas.

2 Messdeck Life

Pictures of messdeck life in all the ships are very similar:

'I joined *Repulse* as a boy seaman and lived, with about 60 other boys, on the boys' messdeck . . . broadside messes, each one of which consisted of a long table with a crockery/bread locker at the ship's side end and long wooden stools either side. Hammocks were slung over one's mess table and, during the day, stowed in hammock netting in a flat below the messdeck; hammock nettings were bins with sides about 3 feet high. In the same flat were the boys' lockers, we each had one in theory; in fact there were never quite enough to go round and, in any case, they were not big enough to hold a kit and we used our kitbags which were stowed in a rack.' *C. T. Piddington*

'The choice of a billet to sling one's hammock depended upon the conditions prevailing at the time . . . Passageways were the places to be avoided when slinging a hammock: crew members making their way to the heads never managed to duck low enough, and occupants of hammocks received violent blows in the back many times during the night. Every inch of space was taken up by prostrate bodies. Mess tables and mess forms were even utilised as beds.' *A. Daniels*

'We never took our clothes off at sea, except odd times when we would sneak a quick wash all over . . . For sleep, any locker, table top or mess stool — one could sleep on anything, even if only about 4 inches wide.' *R. V. Racey*

'In my messdeck, and in many others, the hammocks were often "double-banked"

Right:
The use of hammocks meant that men could sleep in the same areas in which they ate and spent much of their off-watch time, the hammocks being stowed away during the day. *IWM A30162*

Below right:
A messdeck onboard *King George V* at meal time. The food has been brought to each table in trays and served up at the table. *IWM A2858*

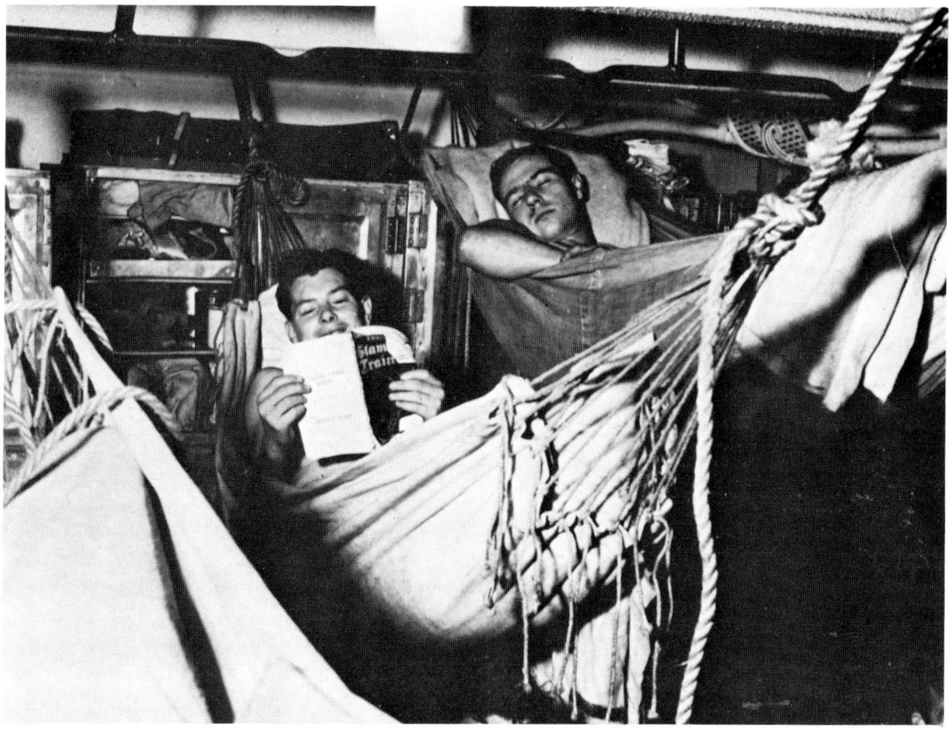

not at all pleasant and very cramped.' *B. C. Cambray*

'Messes had one Leading Rate in charge and anything from 10 to 20 members in the mess . . . Our mess had to be used at sea for eating, writing, sleeping; any relaxation such as knitting, which was done, making wire flowers, drawing — all had to take place in our crowded messes, especially during our first two years on North Atlantic convoys.' *R. V. Racey*

Messes varied from those which contained a large proportion of regulars to others which might be mainly Hostilities Only ratings:
'The mess I was in was a typical cross-section of all walks of life — from bank clerks to barrow boys and, as you may guess, from all parts of the UK — Scots, Irish, Geordies, Brums, Scouses and Londoners.' *E. D. Sharpe*

Despite the wide variety of backgrounds, most fitted in and the mess became their home, and their messmates their friends. There was a strong feeling of camaraderie and of discomforts shared. Your 'oppos' could be trusted:
'Stealing — not once did anything go missing that I knew of, and woe betide anyone who was stupid enough to put a lock on his locker.' *B. C. Cambray*

There was a definite ladder of status on a messdeck. The Leading Hand of the Mess reigned supreme, with the three-badge ratings next in the pecking order (men with three Good Conduct Badges signifying at least 12 years' service). Thereafter, length of time in the mess was an important factor, particularly when it came to finding a billet for your hammock:
'Space was still at a premium, particularly with regard to hammock berths . . . your hammock space was "earned" as time went by and drafts left the ship, spaces being taken in turn by length of service in the ship.' *J. E. Bowen*

Food was a most important consideration in life onboard. The so-called General Messing system was in use in all large ships, whereby all food preparation and cooking was done by the galley and the financial side was also centralised. This may seem the obvious way to cater, but in World War 1 most ships were on 'Broadside' messing: each mess had an allowance out of which they 'bought' their food and prepared it on the messdeck. The galley only had responsibility for the cooking. Hence the term 'cooks of the mess' lingered on for the two men detailed daily to clean the mess and look after such things as the bread issue, tea, milk, etc. They might, however, find themselves involved in peeling the potatoes!

'The general messing arrangement entailed collection of meals from the galley in large trays, 16 rations to each tray and usually 16 ratings to a mess. The daily menu was promulgated along with daily orders, and at night a vat of cocoa was made with big slabs of chocolate and kept hot with a steam jet laid in the vat. This was for the use of watchkeepers.' *G. A. Langridge*

It is said that if sailors are not complaining about the food, then there is something really seriously wrong but, in fact, it would seem there were few complaints:
'The food was reasonable and varied. We never starved — in fact we were better fed at those times than a great deal of people at home.' *R. V. Racey*

B. C. Cambray has even more praise for the food:
'Food. This was truly excellent — plenty of it and lots of variety, although it was noticeable that the Chiefs' and Petty Officers' messes did rather better than the rest of us. The tinned tomatoes were a case

in point — they got tomatoes in juice, we got juice with tomatoes! It is funny how things like that stick in one's mind after so many years. I still think of the delicious smell of the newly baked bread — it may have been white (not the "thing" these days!) but it was superb.'

He did have the advantage in serving in *King George V*, one of the most modern ships. In *Royal Sovereign*, the galley was a little more old-fashioned:
'Also coaled ship — though nothing like the old days. *Sovereign* still had coal galleys, so boys still had the job of coaling.' *F. Troughton*

The Paymaster did not always get it right:
'General messing — all catering being done by the Paymaster and his staff. The food in general was very good, the only dish which few would eat was tripe and onions. It usually went straight from the galley to the gash.' *A. W. R. Brown*

The senior rates had 'messmen' to look after their mess:
'The mess was kept clean by two messmen, probably ordinary seamen. They collected food from the main galley and served to individuals, and cleared it away after meals. The extra money they received for this service was paid once a month from mess bills collected by the mess treasurer.' *W. Burley*

The rum issue was an important event in the day:
'The majority of men entitled to draw took their rum, three parts water, one part rum, total mixture half a pint; this was issued at midday. The basins (we did not have cups) were set out at the end of the mess table. We had a mess ladle which held one pint and, by careful handling, one could measure half a pint. If someone had a birthday he would get "sippers" from the others.' *A. W. R. Brown*

Above:
The contrast between the messdecks of the ratings and the Admiral's quarters: Rear-Adm Mack in his cabin. This, however, would have been the Admiral's harbour cabin; his sea cabin near the bridge would have been more spartan. *IWM A14298*

Left:
The rum issue on board *Rodney*. One man from each mess attended with a 'fanny' in which to collect the issue for his mess. The copper jugs on the deck are the rum measures. The 'Tanky' is pouring one issue into a fanny. *IWM A103*

Right:
Thankfully, not an everyday occurrence — painting ship, in this case *King George V*. It was a major operation with two men on each stage doing a 'fleet' or strip down the side, and then moving aft to do another. *IWM A9463*

The issue of rum actually entailed quite a lot of work (including precautions to make sure it was not abused), and a degree of tradition and ceremonial about it. G. A. Langridge, a supply rating in *Howe*, was frequently involved in the rum issue and has described it in detail. Note that *Howe* ratings received a mixture of two parts of water to one of rum:

'Ration was ⅛ pint per man, neat to Petty Officers and above (excluding officers), and with 2 parts of water to other ratings. At 11am the pipe "Up Spirits" for the neat issue made from the rum store, then at 12am the pipe "Grog issue" at the rum tub. The rum for the 12 o'clock issue was drawn at the time of making the next issue and placed in a special barrico [small barrel]; this was then placed below the quarterdeck where a Royal Marine guard was always stationed at the entrance to the officers' quarters. The rum tub was always positioned in a flat below the upper deck in the waist of the ship.

'During the forenoon, messmen left a mess fanny [container] full of water lined in order in front of the tub. At the issue, the Supply rating, along with the Rough Book which showed the ration to each mess, stood on the left of the tub, a Warrant Officer to the right and, in *Howe's* case, a Royal Marine known as the "Tanky" held the copper gallon measure of the tub. The messmen then proceeded to fill the measure from their mess fannys, the gallons emptied being marked off by the Supply rating until the required amount was in with the rum, also emptied in from the unlocked barrico. The whole lot was given a good stir and issue commenced, each messman calling out "1 Mess, Sir!" or "2 Mess, Sir!" accordingly. The Supply rating would call out the ration, eg if there were 15 drawing rum the call would be "1 out of 2", or 2 pints less 1 tot, so Tanky would put 2 pints in and then take 1 out with a tot measure (all very complicated until you got the hang of it). At the end there had to be only 2 or 3 pints left over. Heaven help the Supply rating if there was, say, a gallon! There would be complaints of a weak issue. [A slight excess of water was always added to make sure that there was enough to cover any slight inaccuracies in measuring out the issue.] The Warrant Officer then instructed them to "See it (the residue) down the scuppers". Rum was a form of currency at times, "Sippers" for small favours, "Gulpers" for a bigger favour, say a loan for a run ashore — anything in fact. A tot could work miracles on the lower deck.'

On the messdecks at 21.00 rounds, the MAA and Commander found inert bodies, particularly in the RMs' and Stokers' messes. It was discovered they were paralytic with neat rum. A member of the RMs, who turned out to be a professional safe cracksman, had broken into the ready-use rum store and stolen several gallon jars, flogging it around the messdecks at one shilling per cup.

The rum Tanky was a three-striper but he never held his stripes very long. He was also the gun sweeper of S2 where any spare rum left over after issue was ditched down the scupper. That was the cleanest drain in the Navy, because in it the Tanky had a polished paint pot that caught a fair drop of 2×1 [diluted rum] that was ditched. Then in the afternoon the Tanky and his mate, another three-striper, would sometimes unscrew the scupper cover and have an extra tot or two.

The routine in harbour worked by the ship's company was similar to the routine used in peacetime although, depending on the threat, a proportion of the armament might be manned. A typical day might be:

'Call the hands was at 05.30 and hands fall in at 06.00. Scrub decks until 07.00, then quarters clean guns until 07.30. Then hands to breakfast and clean into dress of the day. Divisions at 09.00 which might be followed by PT. Then work part of ship with a stand easy at 10.30 for ten minutes when you could smoke. Dinner 12.00. Hands fall in at 13.15 and a stand easy at 14.30. 16.00 hands to tea and change into night clothing — this would be a blue serge suit.

'We had a fairly large recreation space and from time to time we would have a film show. Ships' companies provided their own entertainment: Mah Jong was very popular, also solo whist, cribbage and darts, and in fine weather on deck you would see a group of men sitting playing "Priest of the Parish". There were also a lot of other activities. Most of the 6in casemates would be in use — a choice of two or three barbers, tailoring, suit making, shoe repairs and mat making.'
A. W. R. Brown

At sea the routine depended very much on the theatre of operations and the possibility of attack. We have heard earlier on how some ships early in the war spent many hours at Action Stations. R. V. Racey in *Revenge* found:

'Cruising Stations were used mostly by us whilst at sea. Four hours can be a very long time especially if you are cold, but other times watching the destroyers moving in and out of the convoys somehow seemed to make the hours fly by. In the night watches there was always "Kye" [hot chocolate] and hard biscuits brought round, the Kye almost always thick like soup, but always welcome.

'Defence watches were only used if an enemy of any kind had been reported, but of course Action Stations were many times used at dawn or emergencies, but on the whole, most of our watches were at Cruising Stations. [Broadly speaking, Cruising

Above:
Cruising Stations — one-third of the ship's company closed up. *Repulse* with a troop convoy. *IWM A6793*

Right:
'The upper deck, whenever we possibly could, was holystoned and scrubbed. . .' Men at work on the upper deck of *Resolution*. *T. W. Pope*

Below:
Home for 1,400 men: *King George V* at sea. *IWM A5994*

Stations required one-third of the crew closed up, Defence watches half, and Action Stations required the full crew.]

'There was always work of different kinds to do when not closed up, or even in heavy weather, such as extra cleaning of flats, spaces, messes; painting, chipping, wire splicing, rope splicing, making grommets, heaving lines, maintaining blocks and tackles of many kinds. There were many things to do but, again, after the "duty day" had been done and time was your own, if you were cooped up down below the hours seemed endless. Many just found a billet in any place possible for a kip, and arranged for a call when due on watch! Others would write letters, mend and darn clothes, do pen and pencil drawings, paint, read. Others like myself would do embroidery, or go and visit other shipmates and friends, or go to the fairly large recreation room starboard side under the boatdeck where one could smoke and chatter. Inboard of the same recreation room we had a small "goffer" bar, soft drinks being sold by the NAAFI.'

The Navy has always been particular about cleanliness:

'We were fanatical about cleanliness. The upper deck, whenever we possibly could, was holystoned and scrubbed; the sun bleached the deck white. In the messdeck we had a kind of thick red cork flooring. Each day the mess would be scrubbed out. We took it in turns, down on our hands and knees with a block of "pusser's hard" [soap], a bucket of hot water and a scrubbing brush, and that would be scrubbed immaculately every morning. The mess table, which wa one long white wooden table, was scrubbed We had a "scran bag" onboard. If you lef anything out, tough, because it went into th scran bag — that made everybody tidy an clean.'

The Royal Marines lived in exactly the sam conditions as the other ratings although thei mess was usually separate. Also messin separately from the men were the boys, thei mess was the same as all the others except:

'In one corner of the boys' messdeck ther was an enclosed instructors' mess wher lived three fearsome Petty Officers, ou instructors. Although the war had started

boys were required to continue school and instruction as well as normal ship work and watchkeeping, Action Stations and Air Attack Stations . . . At sea I kept watches as a lookout, and parachute mine lookout in harbour. I think I undertook much more physically hard work as a boy than later as a man because, when extra hands were needed, for carrying stores for example, the captains of the tops [Petty Officers in charge of section of the ship] would send the hands they could best do without — boys! . . . Quite an amount of our duties and instruction brought us in contact with Midshipmen who had to learn the same things. I always felt a sympathy for them because, although their living standard was far above ours, they had more instruction on schoolwork, were subjected to pressure all the time and suffered the cane a great deal more than boys. A boy was caned for a fairly serious offence, one which would send a man to cells for a few days; Midshipmen were beaten for minor offences by the Senior Midshipman, the Sub-Lieutenant of the Gunroom and, in more serious cases, by

Below left:
Divisions, which involved inspection by the Captain, with everyone in their best uniforms. *S. Pearson*

Below:
Not so much a picture of a 16in shell, more an example of the variety of rigs which might be worn, in this case displayed by a fusing party on board *Nelson*. *S. Pearson*

Bottom:
The ship's company who had to be organised into departments, the watchbills made out, specialist parties detailed off, etc. Here, the ship's company of *King George V*. *J. E. Bowen*

Right:
Action Stations, when everyone on board was closed up, in this example in the control top, the highest point of the ship. *IWM A1510*

Below right:
Another position manned — the Action Steering Position in *King George V*. Note how young the boy manning the telegraphs looks. *IMW A1782*

the Master at Arms. Midshipmen and boys parted ways on promotion to Sub-Lieutenant and Ordinary Seamen respectively. Being rated Ordinary Seaman changed my life in a variety of ways. I could stay up until "Pipe Down", stay ashore until 22.00 (midnight on rare occasions) and, best of all, I no longer had to go to school. (We carried two schoolmasters and two Instructor Lieutenants.) My duties changed little; I was given my accrued "credit", the pay for my years as a boy, and received my full pay of twenty-eight shillings a fortnight, riches indeed!' *C. T. Piddington*

Some boys had the task of lashing up two middies' hammocks apiece as well as doing their own, and then getting washed and into breakfast. The middies just wouldn't get out of their hammocks until the last minute, consequently we were often late for breakfast. We complained and the Commander took it up. He came down one morning, sized up the situation and sent us boys on our way.

About an hour later — what a wonderful sight — the middies climbing the mainmast with their hammocks slung over their shoulders; the Commander sent them up higher and higher. The fumes from the funnel were blowing on to them, and there they remained for some considerable time. Joy indeed! No more trouble with them.

Occasionally pets found their way into the messes, such as the kitten in *Nelson*:
'In our No 42 Mess we had a kitten, black and white — heaven knows how we were allowed to keep it, if in fact we were! It was a dear little fellow who really did feel seasickness, or so we thought. So we made a complete replica of a hammock, together with stretchers, nettles, everything — a tiny one to suit the size of the kitten, and slung it under our messdeck table. We taught this kitten to get into the hammock whenever the ship used to lift and fall like a slow lift; one remembers the slow pitch of a battleship. And so there it lay on its back, perfectly happy. Whenever there was sufficient sea to make "Nellie" roll and pitch, this kitten went to its hammock automatically.'

There were other characters on *Nelson's* messdecks as well: 'I well remember the 3-badge ABs rolling their leaf tobacco using spunyarn wound round and stretched taut by sitting on the line, between hammock bars. They added rum to it and then the "prick" of tobacco was allowed to mature, before being cut up to be chewed.' *J. V. Haddock*

Men took great pride in their messdecks, never more so than when they were inspected during Captain's Rounds. John Haddock recalls:
'Messdeck rounds on Saturdays was always a very important inspection by the Captain. We polished everything in sight and polished it again. Then there was a very important way to lay out all the wares on the tables and each Leading Seaman tried to make sure his was better than the next table. We had some intriguing patterns of cutlery which were amazingly artistic and beautiful. Everyone took a great pride in his mess, the cleanliness and being tidy.'

The organisation of a ship as large as a battleship was a complex affair. The

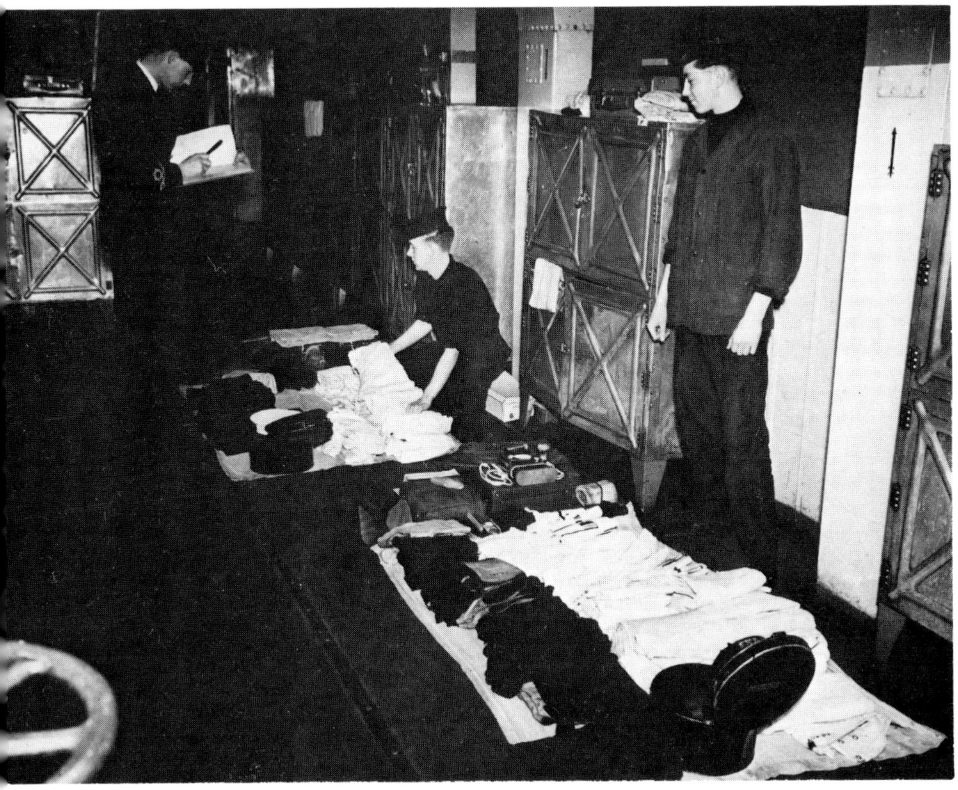

Left:
Another important part of the ship's organisation was the Divisional system, where each man had an officer who was responsible for his welfare, training for advancement, etc. In this photograph, the Divisional Officer is inspecting the kit of some of his men. The kit laid out on the deck would be a man's entire set of uniform clothing. The lockers in which the kit was stowed can be seen behind the men.
IWM A2215

Below left:
Something out of the normal routine: a visit by the Prime Minister, Winston Churchill, and his wife to the Fleet at Scapa Flow. *W. H. Smith*

Commander was responsible for the overall organisation and administration of the ship and was the Head of the Seaman Dept. Each department — Seaman, Engineering, Supply etc — had a Head of Dept, and then under him would be more junior officers, each with his own responsibilities. Then the management chain went down through the Chief Petty Officers, Petty Officers and Leading Hands to the seamen, stokers, radio operators, supply assistants, sick berth attendants, etc, who made up the bulk of the ship's company, which might total some 2,000 men.

Each man had his own place in the organisation, he had a mess in which to live. He was also a member of a Division, with a Divisional Officer who was responsible for reporting on a man's suitability for rating up (promotion), encouraging men to qualify for a higher rate, looking after his men's welfare and generally taking an interest in their affairs and being someone the ratings could turn to for advice and guidance.

Each man also had his own position to fill at Action Stations, Defence Stations, Cruising Stations, entering harbour, leaving harbour, etc. He was also a member of a watch, so that when a watchkeeping routine, such as Cruising Stations, was being worked, he knew when he was required on duty. He also had a 'part of ship' — a place of work when he was not on watch. A seaman might be a fo'c'sle man, responsible for the cleanliness, maintenance and painting of the forward part of the ship. A stoker might be detailed for a particular boiler room. Quite a lot of men belonged to a 'party': they were special teams who looked after particular equipments or small sections of the ship such as boats, double bottoms, food stores and so on. Ratings also, of course, had their own trades from seaman gunner to stoker, cook to signalman, ordnance artificer to engine room artificer, steward to sick-berth attendant. So that each man knew exactly his stations and jobs, a 'Watch and Quarter Bill' was made out which listed all the men and their positions at Action Stations, Cruising Stations, etc.

Up on the boat deck was a cubby hole where an old (to me) three-badge AB spent his time producing the most magnificent ropework and knots for the Admiral's and Captain's barges. A kindly soul, he taught me a tremendous amount, not only about knots, but all those little wrinkles that seem to come only with experience and a real love of the Service. Some of his work was truly remarkable — so remarkable that Capt Mack and Adm Tovey used to spend the odd quarter of an hour or so just watching and enjoying his craft.

There was an enormous amount of gambling, men losing their pay within a few hours of getting it. The usual game was brag — poker never reached the messdecks, which was just as well. We had a chap who had been a magician before the war — he would never touch the cards for if he won. . . He showed us why and how: with a new deck he always dealt himself winning hands. With us all clustered round, he would play 'Find the Lady' and no one could ever see where she was.

3 Damage Control

Battleships were designed to withstand damage; they had armour plating protecting the vitals of the ship. An armoured deck ran for much of the length of the ship, not at upper deck level but one or two decks below the upper deck. An armoured belt protected the side, along with a complex arrangement which might include bulges or compartments built into the side of the ship specifically designed to absorb the impact and explosion of a torpedo. Gun turrets were armoured, as was the conning tower and armoured gun control tower.

Not only did the ships have various forms of protection but they were also sub-divided into many compartments: each of these could be isolated by shutting watertight doors, hatches and ventilation valves so that if one part of the ship was flooded, it would not spread. Similarly, fires or the effects of shells exploding inside the ship could be contained.

Many doors and hatches were kept shut except when required for access; others remained open for normal steaming but would be shut at Action Stations. Damage Control routines and precautions were part of the way of life but were often inconvenient:

'Damage Control was first class. Everyone was most conscious of the necessity of operating watertight doors and ventilation trunking valves in various states of readiness. It was difficult getting around the ship at sea, especially with the high welded plates across all the doorways. These would temporarily hold back any initial rush of water from one compartment to another, enabling the watertight doors to be closed.' *F. Troughton*

'Whenever you walked from the messdeck aft to the heads forward, you had to walk a criss-cross pattern as alternate bulkhead doors were closed and cleated up, to make closing up quicker and safer against an inrush of water.

'At sea, every time it was necessary to open a hatch and leave it open for any length of time, you obtained a metal label which was issued from the Damage Control Centre. It was hung at the top of the hatch to indicate someone was below; it was returned immediately you had closed the hatch. Thus, at any time, hazards could be seen at a glance. If action was imminent, pipes would be made to alert all below and closing of all hatches was carried out.' *G. A. Langridge*

Armoured hatches which penetrated the armoured deck were a particular problem:

'The hatch was approximately 9in thick. It must have weighed many tons and, although in my physical prime at 23, I remember how taxing I found the effort of raising this hatch with the chain block attached. I used to slide underneath when the hatch had just been raised enough.' *N. Hill*

'Armoured hatches were a hazard in harbour, but could be even worse at sea when one waited for the roll of the ship to help bias the weight in favour of either closing or opening the hatches, which were actuated by massive springs. There were several nasty accidents to feet and hands.' *W. Burley*

The armoured hatches could only be operated from above, so when men had to go to their Action Stations in positions below the armoured deck. . .:

'The final thump as the armoured hatch was lowered and the clips put on — I realised the clips could only be opened from topside; below we couldn't open them. I felt trapped and afraid.' *E. R. Moyle*

Damage Control parties were trained to deal with the effects of bomb, shell or torpedo damage. They knew their sections of the ship in detail — where all the valves, fire hydrants, hoses, emergency lights, alternative electrical supplies, etc, were to be found. They could carry out temporary repairs, such as shoring up a bulkhead or plugging splinter holes. They had breathing apparatus for working in smoke. Damage Control activities were controlled from the Damage Control Headquarters, known as HQ1. It was here that more drastic actions might be decided upon, such as flooding a magazine which was threatened by fire, or counter-flooding part of the ship to correct a list.

The full damage control organisation closed up at Action Stations; at more relaxed states, there was always a skeleton organisation. For example, patrols went round the ship checking that rules were being obeyed for watertight doors and hatches, and it was they who would start shutting the ship down if Action Stations was ordered:

'In action I was part of a Damage Control team. At Action Stations in a big ship — there are so many group functions — ours was but one. At the pipe "Action Stations" plus the bugle call, as part of the damage control team we had to secure all doors and hatches. Dependent on the type of action — air, sea or bombardment — our stations varied in location to suit a damage control programme. I found that when in a team and the ship had received hits from shellfire or bombs, the thought of losing your life did not occur.' *C. B. Elsmore*

Right:
The men of a 6in gun turret resting at their Action Stations. One man on the left remains alert and mans the communications.
IWM A2134

4 Rough Weather

Left:
***Queen Elizabeth* taking it green in what appears to be a moderate sea. These ships were very wet below decks forward in rough weather.** *IWM A19043*

For the Home Fleet the middle years of the war meant the continued protection of the convoys from German raiders. We will have a look at this phase of the war from the German viewpoint shortly. For the British it meant long periods at sea and, in the winter, it was not just the Germans who were the enemy but the weather as well.

'All the winter crossings of the Atlantic were very cold on the messdecks. No one was allowed to undress or sleep in hammocks, and if the weather was rough, which it often was, the forward messdecks were running with water. We slept on tables, seats, lockers, any flats which did not have water in them.

'The ship was a good ship in very many ways [*Revenge*]. She would roll, dip and ride the waves like a roller-coaster. My Action Station was on the "iron deck" and in many gales I have watched the bows sink down-under as far back as B turret, A turret being completely under water; then she would recover, shake off the water like from a duck's back and ride on top of the sea again, again and again. Sometimes it would go on for days on end. If you were down in the messdecks trying to walk forward to the heads, your stomach would feel like it was dropping out when she was on the downward plunge, then when she was rising, you felt like you were being compressed.' *R. V. Racey*

'At sea in rough weather always meant plenty of water inboard, particularly in the main heads which were right forward, and in the 6in gun batteries which were situated on either side of the ship and which had a very low freeboard. A heavy swell would ensure the bows dipped considerably and a huge cascade of water was thrown back, sometimes over the forward A and B turrets. The quarterdeck would also invariably be awash, but below decks, even in very rough weather, life was never too badly affected.' *E. Taylor*

Resolution had her own idiosyncrasies in rough weather: 'The ship was very comfortable except in very bad weather. In heavy seas she had a most peculiar motion: her fore end would rear right up, then she would plunge to starboard with a corkscrew motion.' *A. W. R. Brown*

Living in a ship in rough weather is very tiring and, in winter, the cold also sapped one's strength:

'I recall huge seas and gales, the cold, bitter cold; on watch with the spray freezing on all the upper works, the steam pipes, hammers; sledge hammers being used when the seas abated to try to get the ice off.

'I recall we were always tired, very tired. We seemed always to be on watch or sleeping, but morale was high. Jerry wasn't going to beat us.' *R. V. Racey*

'A veteran warship of World War 1, *Queen Elizabeth* frequently showed signs of her age by shipping vast quantities of water whilst battling against heavy seas, and a good deal of this found its way into the fo'c'sle messdeck, entailing major mopping-up operations.'

5 German Operations

And what of Jerry? What was life like for the sailors in the German battleships? The Germans had started the war with only four battleships operational: there was no German High Seas Fleet in this war. Of the three pocket battleships, only *Deutschland* and *Graf Spee* were available in 1939; *Admiral Scheer* was undergoing major machinery repairs which kept her out of action until September 1940. We have already heard of the fate of *Graf Spee*. *Scharnhorst* and *Gneisenau* became a famous pair of ships, frequently operating together in the early years of the war and causing significant losses to the British,

Right:
When she was commissioned, *Bismarck* immediately became the biggest problem for the British Home Fleet. She posed a significant threat to the convoy routes and would be a formidable opponent, armed with 8×15in guns and with the most modern armour. She was also fast, being capable of 29kt. *Real Photos*

Above:
***Admiral Scheer*, a sister of the *Graf Spee*, carried out what was possibly the most successful sortie by a German ship against the convoy routes.** *Real Photos*

Right:
The after turrets of *Tirpitz*. Note the gun director in the lower part of the photograph which is different from the British type. In German directors, the optics were on top but the sighting was done through a periscopic system from underneath, as was the rangefinding. In German nomenclature, these were 'Caesar' and 'Dora' turrets. *IWM MH7129*

particularly in the Norwegian campaign. Only two more battleships would join the German Navy in World War 2: *Bismarck* commenced her sea trials in 1940 but was not operational until early 1941. *Tirpitz* was not readly until the end of 1941.

All these ships spent much of their time in harbour. The Germans had no convoys which needed protection and their large battleships were to play no part in operations other than in the North Atlantic, Norwegian and Barents Seas. After the successful invasion of Norway, the Germans had the many fiords of the Norwegian coast at their disposal in which to hide their ships. From these fiords they could slip out surreptitiously to prey upon the merchant shipping which was the lifeblood of the British Isles.

The sailors of the two navies, British and German, would find many similarities in their life styles. When the German ships were back in the German ports all was well, but when they found themselves in the Norwegian fiords, life was very much like that in Scapa: there was little leave and nowhere to go. Men had to make their own entertainment, and the entertainments were very similar to the British. But, because the German ships did not spend as much time at sea, they were not as hardened to the rigours of rough weather, cold and tiredness as their British counterparts.

The sailors would have found their way around each other's ships without difficulty. The German vessels were handsome ships; *Scharnhorst* and *Gneisenau*, *Bismarck* and

Left:
***Tirpitz* seen from *Prinz Eugen* after her return from Spitzbergen. *Tirpitz* took part in few operations, but her mere presence in Norwegian waters tied up units of the Home Fleet for many months.** *Real Photos*

Tirpitz were very similar in appearance, the latter two scaled-up versions of the former in many ways. They had long, low fo'c'sles rising in a handsome sheer to a raked bow; nevertheless, they were still wet forward in heavy weather. They were all fast, capable of 31 and 29kt respectively.

Some technical details varied between the two nation's ships. For example, German gun directors consisted of a rotating rangefinding cupola on top but the sighting was done from beneath through a periscope. In British directors, sighting and rangefinding were done from the same position high up in the director. Some nomenclature differed: British turrets were always referred to as A, B, X and Y; the Germans used Anton, Bruno, Caesar and Dora. Each German ship was divided into 12 Divisions:

1 to 4 — main and secondary armament
5 & 6 — anti-aircraft guns
7 — service department (ie cooks, stores, etc)
8 — ordnancemen
9 — communications and quartermasters
10 to 12 — propulsion

In October 1940 *Admiral Scheer* sailed from Germany on what was to become possibly the most successful sortie by a German capital ship. She broke out into the Atlantic through the Denmark Strait in November and commenced operations against Allied convoys. Using her aircraft for reconnaissance, a convoy was spotted three days later. One lone merchant ship was encountered and sunk whilst *Scheer* closed the convoy. However, the intervention of the Armed Merchant Cruiser *Jervis Bay* in a courageous, but hopeless, one-sided action did allow the convoy to scatter, whilst *Jervis Bay* herself was sunk. Despite the lack of success against that particular convoy, the presence of *Scheer* in the North Atlantic was enough to disrupt the convoy system for a significant period.

Scheer was to remain at sea for over five months, supported by a system of supply ships; she operated as far as the Indian Ocean. Her tactics were very similar to those used by the *Graf Spee*, approaching unsuspecting lone merchant ships end-on to avoid recognition until the last moment. In common with *Graf Spee's* practice, merchant crews were allowed off their ships before they were sunk. She sank 16 ships plus *Jervis Bay* before returning to Germany at the beginning of April 1941. Although her score was not large, her mere presence had been enough to hamper convoy operations and to force the British Admiralty to devote even more of its scarce resources to the protection of convoys.

Whilst *Scheer* was still at sea, *Scharnhorst* and *Gneisenau* left Kiel in January 1941 for a sortie into the Atlantic. Their operations were confined to the North Atlantic under the command of Adm Lutjens. The first convoy they came across on the Halifax route looked easy pickings until they sighted the fighting top of an 'R'-class battleship; this was *Ramillies*, whose presence was enough to make the German Admiral break off and retire, perhaps wisely. *Ramillies*, old though she was, could outrange both German ships.

For those in *Ramillies*, the spotting of the German ships was the culmination of long months of lookout duty:

'Many bitterly cold days [in the 15in spotting top] — windows down, watching horizons; green seas coming over the bows. Woe betide anyone who took his eyes off his binoculars: a hefty boot soon reminded him where his duties lay. It was from up here that we spotted *Scharnhorst* and *Gneisenau*.'
W. E. Broomfield

Later, in early March, the presence of another British battleship protecting a convoy again put Lutjens off attacking. *Malaya* was escorting a convoy homeward bound about 350 miles north of Cape Verde Islands, when her aircraft sighted the German ships. *Malaya* stayed with the convoy — she did not have the speed to catch the German ships anyway — and the Germans did not attack. *Scharnhorst* and *Gneisenau* entered Brest in late March, having sunk or captured 22 ships in their two months at sea, but having been put off from even greater damage by the presence of British battleships with convoys on at least two occasions.

There was now a short lull in German operations before *Bismarck* sailed on her first operation; but for the British, the protection of the Atlantic shipping continued without a break.

6 *Bismarck*

Bismarck, in company with the cruiser *Prinz Eugen*, sailed from Gdynia on 18 May 1941. The story of her sinking of *Hood*, the subsequent search by British warships and her final destruction has been told many times. However, the impressions of some of the men who were there are worth recording.

When *Hood* and *Prince of Wales* intercepted *Bismarck* south of Iceland, it is well known that *Hood* was lost along with all but three of her ship's company. *Prince of Wales* was fresh from building and suffered from technical problems with her main armament, although she was able to obtain two 14in hits on *Bismarck* before having to break off. For A. V. Godding, who was an Ordnance Artificer in *Prince of Wales*, the technical problems were considerable:

'We were accompanying *Hood* to Iceland when we were diverted to intercept *Bismarck* and *Prinz Eugen*. We closed up at Action Stations and our first action after loading the guns was when the enemy was at about 22,000 yards, and A and B turrets opened fire. Owing to the ship's position at this time, Y turret could not bear on the enemy and he held fire until we were about 16,000 yards range. I remember the range because just prior to opening fire, the Royal Marine on the local control rangefinder had called out "Range 16,000 yards!". I was at this time in the gun house.

'Now came the breakdown, which some "historians" say "the turret would not train"! The cause was that the anti-surge stops on the shell ring sheared and a shell surged back, jamming the intermediate shell trays, which were the links between the shell rooms and the shell ring. The gun house was told "Stop Training!" because of the mayhem being caused in the shell handing room. All hands were then at it with sledge hammers and hacksaws but it was sometime before we were back in reasonable shape on the shell ring.

'Now the central hoist cage had problems. The shell arresters in the central hoist cages were sluggish in returning and the cages were being hoisted before the arresters were clear, hence another breakdown. Cordite handing room interlock 14 was causing problems with the indicator in the traversing space. Incidentally, I modified this interlock to prevent this problem.'

Rodney and *King George V* finally engaged *Bismarck* on the morning of 27 May after she had been damaged by an aircraft torpedo attack. Two accounts from men who manned the armament on *Rodney* tell of their experiences:

'Then came the sinking of the *Bismarck*, actually on my birthday; what an impressive sight, but fearsome nevertheless. We first opened fire around 9.30. She replied and the Gunnery Officer said she was 30 yards short. The next went through the upper rigging and we all prayed, I can tell you . . . I was a gun-layer on the 6in turret and the small aperture in front of me gave me an excellent view, the sight of which, after seeing men wearing white flags, or anything white, and jumping overboard, I'll never forget.

'We were the only battleship to have ever fired 21in torpedoes (six in all), and they all missed, so you can see how close our encounter was. A full 16in broadside was also fired while steaming at full speed. Naturally the ship was internally damaged. Firing the broadside virtually stopped the ship, even though she was running at full speed. The messdecks and most of the fittings were damaged, fan shafts were twisted out of shape, messdeck gear scattered all over the decks, lockers split open. To see such a sight was unbelievable.' *W. Handley*

'During the action I heard the guns banging away — the 6in, the thunder of the 16in guns and I felt the ship shuddering. I heard clangs and thumps and the sound of water being torn apart. It was unlike the sound of depth charges exploding, which I had learned to live with when serving aboard a frigate in later years. When we shouted up to the gun turret, we were always assured that "It's alright, it's ours". I still wonder.

'As the 6in shells came up the hoist I lifted them out, moved them across and put them into the hoist up to the guns. Before the action commenced, we just lay around the handing room trying to relax — it wasn't easy when you are just a lad and scared as hell. It was like a dream, a world of fantasy: am I really here? Is this really happening to me? We wrote on the 6in shells with chalk before they went up to the guns. We sang and chatted away — nerves. At last it was all over and the hatches were raised. I climbed and ran and, at last, I had room to move about, and space; but on the messdecks it was chaos.' *E. R. Moyle*

For the men in *Bismarck*, what had been only three days earlier a magnificent adventure when they had sunk *Hood* and beaten off *Prince of Wales*, turned into a nightmare. They knew that the British would do everything possible to catch them and, when the Swordfish attack late on the evening of 26 May severely damaged the steering gear, such that the ship could no longer be controlled, they knew that their fate was probably sealed. When *King George V* and *Rodney* came over the horizon on the morning of the 27th, there could have been little doubting. Many of the crew of *Bismarck* were very young and had little or no experience of the sea. *Bismarck* had largely worked-up in protected waters and the men had never seen the Atlantic before. Nevertheless, they fought well and, as ever, the German gunnery was accurate as the above accounts show, although *Rodney* was able to escape any direct hits. However, it is

Left:
When *Bismarck* was detected in Norwegian waters, virtually every large ship of the British Home Fleet became involved, one way or another, in the hunt for her. *Real Photos*

Above:
***Hood* had the guns and the speed to take on *Bismarck* but her age was against her. Her armoured protection had not been brought up to modern standards and the plunging fire from *Bismarck* sank her.** *IWM DS595/18*

Right:
This famous photograph shows *Bismarck* engaging *Hood*, taken from *Prinz Eugen*. *IWM HU381*

Above:
During the engagement with *Hood* and *Prince of Wales, Bismarck* was damaged in the bows. *IWM HU400*

Right:
***Rodney* firing at *Bismarck* in the final engagement. The smoke on the horizon marks the position of *Bismarck* and gives a good idea of the distances over which salvoes were exchanged when battleship took on battleship.** *IWM MH15931*

said that morale was lost in an instant when Adm Lutjens told the men that it was 'Victory or Death!', which is no reflection on the men themselves who, like the British counterparts, were young and afraid — 'just a lad and scared as hell'.

7 Gunnery

A battleship was basically a floating gun platform. The main armament was designed to inflict maximum damage upon the enemy; secondary armament was largely for the protection of the battleship herself. In early years that protection was required against destroyer torpedo attack but, in World War 2, a large variety of secondary armament was fitted as protection against aircraft. 'Everything seemed to revolve around the ship's gunnery and its efficiency.' *F. Troughton*

Let us now look at battleship gunnery in more detail. A. V. Godding describes the 14in main armament in the 'King George V' class:

'Two types of 14in turret, the Quad and the Twin. A and Y turrets were Quad while B turret was a twin. All turrets had local control. Cordite was fed into the Cordite Handing Room through a flashtight hopper and was handled on to waiting trays by cooks and writers in A and B turrets, and by Royal Marine musicians in Y turret. Full charge of cordite consisted of 4×85lb silk bags of cordite, with two having an igniter pad (marked in red) on one end. These charges were then placed in the cordite hopper on the revolving platform and rammed into the Central Hoist Cage.

'Shells in the shell rooms were stored in bins and were lifted by hydraulic grabs and lowered into the shell trays. The flash doors through to the Shell Handing Room were lifted hydraulically and shells could then be rammed on to the shell ring, provided the shell ring was locked to the Fixed Structure Control Position (nicknamed the Piano). This Piano controlled the shell ring from the Shell Room and revolved independently of the turret. Control of the shell ring could then be passed to the revolving turret by disconnecting control from the Piano and connecting it to the revolving structure. The shell could then be rammed into the Central Hoist Cage and IF cordite had also been rammed into the Central Hoist Cage, then the indicator in the Traversing Compartment would show "Central Hoist Cage Ready".

'Raising the Central Hoist Cage would bring the shell and cordite up to the Traversing Space . . . 4 cradles were then used to increase the distance between the centres of shell and cordite until they were in line with the cages in the gun wells. Ramming of shell and cordite into the gun cages was then carried out and an indicator for each position in the gun house showed "Gun Cage Ready".

'Gun cage could now be raised and ramming took place. This necessitated the

Above:
Loading cordite charges into the hoist to go up to the turret, in this case in an American 16in mounting on board *Alabama* (BB-60). *US Navy*

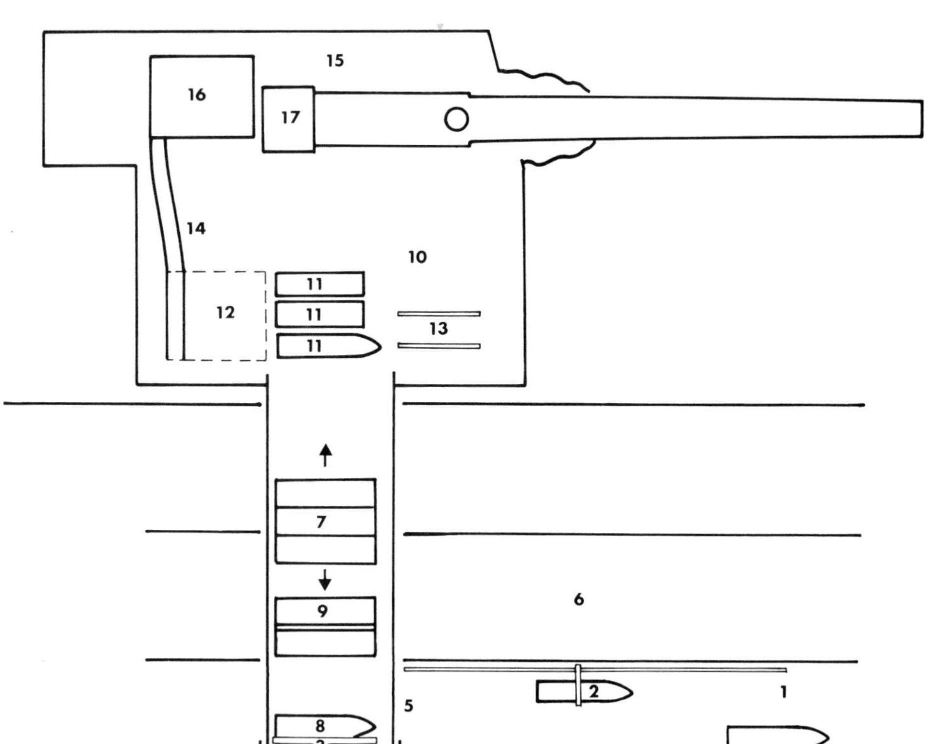

Loading Arrangements — Main Armament

Detailed arrangements varied from class to class but this Figure shows the general principles upon which all turrets worked.

Shells were stored in the Shell Room (**1**) and moved by hydraulic grabs (**2**). They were then positioned on a bogie (**3**) which could either be locked to the Fixed Structure (**4**) or the Revolving Structure (**5**) under the control of the Fixed Structure Control Position ('piano'). Cordite charges were stored in the Magazine (**6**) and placed in hoppers on the Revolving Structure. The Central Hoist Cage (**7**) would descend to Magazine and Shell Room level to be loaded with the waiting shell (**8**) and cordite (**9**). The Central Hoist Cage then ascended to the Traversing Space or Working Chamber (**10**). Here the shell and cordite would be positioned on Cradles or Waiting Trays (**11**).

When the Gun Loading Cage descended (**12**), the shell and cordite would be rammed by hydraulic rammers (**13**) and then the Cage ascended the Guide Rails (**14**) into the Gun House (**15**). Once in the loading position (**16**), the Breech (**17**) would open and the shell, followed by the cordite would be rammed home.

Above left:
Moving a 16in shell on to the hoist: just how large these shells are is readily apparent. This is also aboard *Alabama*. *US Navy*

Above:
The training engine space beneath a British 16in turret. *IWM A15457*

Left:
Gun-Loading Arrangements — Main Armament (not to scale).

cage stopping at 3 positions, first to align the shell with the breech, then ram the shell. The next two positions aligned the cordite and rammed two quarter charges each time. The last two quarter charges had to be the two with the igniters. This then reflects back down to the man who loaded the charges deep down in the Cordite Handing Room, because once loaded in cages, they could not be turned round.'

The above account shows just how complex the guns were. Much of the complexity was due to the requirement to transfer shells and cordite charges from the fixed structure of the ship to the revolving structure of the turret and, at the same time, lift them up through many decks from the shell rooms and magazines deep in the ship to the guns themselves. In addition, flashtight doors were built into the path of the shells and charges to ensure that a hit on the turret did not cause a chain of explosions to run down to the magazines. In the magazines, the cordite charges were heavy to move around: 'My Action Station in Y magazine consisted of loading 80lb charges of cordite through an anti-flash chute to the main guns. As a team, one man opened the canisters with a special tool, another pulled on the tape to withdraw the charge assisted by another. The tapes were removed and it was passed along finally to the Corporal who operated the anti-flash

Left:
Maintenance work in progress inside a 16in turret. *IWM A2123*

Below left:
Sponging out 'B' turret. *A. Daniels*

over the side, then topping the surrounding lockers up with more ammo.

'I was always thankful for the job I was doing and that I was out in the open where you could see and hear what was going on, as opposed to being below decks. Our eight barrels firing together made a hell of a noise, but to me the worst was the 4.7in guns when they went off. I used to stuff my ears with cotton wool and put the ear pads on; it was like the crack of a whip. The 6in and 16in made more of a dull, muffled rumbling sound.' *S. Pearson*

Maintenance of the guns was in the hands of the Ordnance Artificers. A. V. Godding describes some of the problems he encountered on 14in turrets:

'The Centre Pivot Space was one of the biggest problems. There was little head room under the shell room deck plating, about 3 feet, and invariably it was running in hydraulic water. The pipe connections, of which there were many, were always in need of tightening and/or the replacing of joints. Lifting cylinders in the shell rooms used with the shell grabs were also continuously leaking, as were the pipe joints; this caused the shell bins to be always awash with water. The pipes to the shell room flash-doors were very inaccessible between the shell trays, thus also adding to the flooding of the shell bins.'

Some jobs on the guns turned out to be major evolutions:

'We certainly had quite a problem when it was decided to change the inner "A" tube of the 4.5in HA guns. We had 10 twin-mounted turrets and a set of spare barrels were stowed on the hangar deck. These turrets were countersunk so that only a portion of the gun shield was above the upper deck. So as to enable the barrels to be removed, a segment of the gun shield had also to be removed, and this could only be done by drilling out the rivets: they were probably ¾in in diameter. It was a really tough job to be carried out with hand tools . . . With the section of the gun shield removed, a slide was fabricated to support the whole barrel complete with the breech ring, and run it out on to the deck. Then after driving off the breech ring to separate it from the outer cover of the gun barrel, we could now remove the other retaining rings and slide out the old inner "A" tube, replacing it with a new one. Reversing of the process was a lot easier. I think the whole job took about 3 spells in harbour.'

Following a main armament sub-calibre practice shoot off Ceylon, the Ordnance Artificers (OAs) were struggling to get the sub-calibre guns back into their stowage positions. The OA of X turret got a little excited and, standing on the quarterdeck, shouted to the crew above to 'Lower the B. . . . Thing!'. Capt Norman, coming from behind, tapped him on the shoulder, saying 'Handsomely is the word man, handsomely,' in a kindly way.

The amount of gunnery practice varied considerably from ship to ship. In *King George V*:

'Something which was of concern to me then and since was the incredible lack of gunnery practice we had. In the nine months I was aboard, the 14in guns were fired but once and the 5.25in once or twice, as were the pom-poms. I felt that if we came up against anything heavy, we would be hard-pressed to put up a good fight, although she did fire with extreme accuracy at the *Bismarck*.' *B. C. Cambray*

Perhaps this period in *KGV* was an exception, for other ships certainly did practice gunnery frequently. *Valiant*, for example, practised in that area well known to the Grand Fleet, the Pentland Firth, and showed that accidents could happen:

'Early in the morning, at 06.14, *Valiant* left her anchorage in Scapa Flow and steamed into the Pentland Firth. A gunnery exercise had been arranged to carry out a quick fire low-angled practice shoot against E-boats. Practice ammunition was to be used (this is the same size calibre shell but it has a reduced charge and the shell head is filled with sand). *Valiant* was on the range area and at 10.27 the shoot started. The starboard side 4.5in batteries commenced quick firing. At 11.55, whilst the port side 4.5in batteries were firing on a forward bearing, P2 4.5in twin turret fired one shot through P1 4.5in twin turret. P2 turret had overridden the safety training stops . . . a fault in the safety mechanism. This tragic accident killed one man and injured three more, one of whom died shortly afterwards from his wounds.' *J. G. Barlow*

Action Stations could be ordered at any time without prior warning. In addition, it was customary to go to Action Stations at dawn in case the daylight revealed an enemy ship

Closing up to Dawn Action Stations could be done in an orderly manner, the men having prior notice, but closing up unexpectedly, particularly at night, was very different:

If you have never been onboard a battleship when Action Stations are sounded and seen men piling out of their hammocks in the middle of the night, then you've seen nothing, especially in the most northern climes when sub-zero temperatures are normal, with ice an inch thick on the bulkheads. Ratings and Marines pulling on the most varied of garb to keep warm; ratings landing on each other's backs as they race to their respective gun turrets; shouts of "who's pinched my socks?", "where's my anti-flash gear?", not knowing that they had already slept in them.' *W. Handley*

'When Action Stations sounded off, it seemed as though the whole ship was alive. Everyone knew their station from practice and made for it as fast as possible. In my case, going from the Issue Room forward to the hatch aft leading down to the magazine took some little time, and I usually had to unclip and reseal all the hatches leading down and into the magazine.'

As well as the large number of men required to man the guns, there were many other positions to be manned at Action Stations. Some were not very pleasant:

'But my Action Station as a torpedoman on a battleship was not so interesting. I was stuck down in the Diesel Dynamo Room or the Breaker Rooms, the coldest places in the ship on Russian convoys.' *R. Emmington*

'I had various Action Stations, mainly for experience. At one time I was searchlight crew, another I was alone in a compartment for emergency electrical supplies, ie if the compartment either side of me had lost its supply, I would then re-connect it to whichever side had lost it.'
R. G. Duquemin

Above left:
Work being carried out on the 6in turrets on board *Nelson*. *IWM A95*

Above:
A 5.25in practice shoot on board a 'King George V' class vessel. These turrets were designed for both the surface and anti-aircraft roles; anti-aircraft firing could be carried out against a sleeve target towed by an aircraft. *IWM A3921*

Below:
***Anson* carrying out a practice firing of the 14in. The large flag at the foremast, which was red, indicates that it is a practice firing.** *MPL*

Providing food for men closed up for long periods at Action Stations was necessary:

'When at full Action Stations, action messing was put into effect. As "knife and fork" meals were not possible, suitable meals were provided that could be eaten at one's station, wherever this might be.'
J. E. Bowen

8 Russian Convoys

The nature of the war for the Home Fleet changed significantly after the German invasion of Russia in June 1941. The first convoys to Russia started in August of that year and the battleships of the Home Fleet were involved in operations to cover the passage of the convoys; the focus of attention shifted somewhat from the transatlantic convoy route to the waters of the Norwegian and Barents Seas. The threat came, as ever, from the German capital ships. By the autumn of 1941, *Tirpitz* was assessed as being ready for operations and was considered the greatest threat. *Bismarck* had been eliminated earlier in the year. *Scharnhorst* and *Gneisenau* were in Brest. They were to make their famous 'Channel Dash' back to Kiel in February 1942. Of the pocket battleships, *Scheer* and *Lutzow* (ex-*Deutschland*) were operational.

Russian convoy protection would remain a major task for the British battleships until the final elimination of *Tirpitz* in late 1944. However, as is well known, the threat to these convoys came not only from German capital ships, but also from U-boats and aircraft, and also from the weather.

'Chipping parties was the worst chore in the Arctic Circle, going along chipping ice off the superstructure, guard rails and anchor chains. 4-hour watches round the clock, gun turrets always on the move to stop hydraulics from freezing. Every time a wave came onboard, it froze solid.' *W. Handley*

Various fiords on the coast of Iceland were used as bases for Home Fleet ships:

'Serving aboard HMS *Howe* we left the port of Northern Iceland, which at the time was frozen over to about 6 inches of ice, and when we moved the whole sheet of ice broke up. On joining this particular convoy, the weather was so bad we never had one item of trouble from the enemy as the sea was so mountainous, and you often wondered how anyone could survive. I remember going forward on several occasions, and right underneath the bows it was just like someone hitting it with a terrific big sledge-hammer every time a wave hit the deck above. Also on this particular convoy we learned that several merchant ships had to turn back, and one of our cruisers had the top of its gun turret (obviously the forward one) ripped off with the rough seas.' *N. H. Bissell*

After the United States entered the war their ships also assisted in escorting and covering Russian convoys:

The time at sea was generally spent in acting as a covering force for convoys proceeding to and from North Russia. We, as C-in-C Home Fleet [*King George V*], would be fleet No 1 in single line ahead with the *Victorious* as fleet No 2, and two or more capital ships astern of her, a close destroyer screen and sometimes a cruiser stationed well out on each beam. Quite often American capital ships and their destroyers would form part of the covering force. At times German shadow aircraft would locate the fleet, but only once did I see one of these aircraft drop a bomb, and then that was just ahead of the *Victorious* — which did cause some concern.

'On one occasion whilst out on patrol, the ship's company were informed that it was believed a German capital ship was somewhere in the area and could be the *Tirpitz*. I remember how we were closed-up for full Action Stations for most of the day, not quite knowing how it was all going to end. As it happened, only a radar contact was obtained and the ship eventually reverted back to a lesser degree of readiness. On the lighter side, I recall that plenty of cold rations were available with lots of hot soup.'
G. Whyte

There were some lighter moments:
'Towards the end of 1942 we acted as cover for a Russian convoy, keeping well to the west as far as I could make out. We ran into quite a storm with an extremely long swell, so much so that, while we [*KGV*] were crashing up and down, our escorting destroyers were literally riding up and down the seas, shipping very little water — at least that was how it appeared. We came across an armed trawler miles from anywhere on the unpleasant Northern Patrol. The Admiral had a word with the Captain who sent for the Navigator — the idea was to give the trawler its exact position. This was duly sent and acknowledged. After a few minutes, the trawler called us up and started flashing, and I started to chuckle. The Officer of the Watch asked me what the joke was — I suggested that the Yeoman should give him the signal, but he insisted — as a Leading Telegraphist I could read the Aldis as fast or faster than most Signalmen. The trawler's skipper replied to the effect that he

Below:
Howe at sea in Arctic waters; ice is starting to accumulate on the forward turrets.
IWM A15430

Below right:
Ice being cleared from the forecastle of *King George V*. It was important to clear the ice as the extra weight on deck could be considerable and lead to loss of stability. B. C. Cambray

Bottom right:
Ice accumulation on the barrels of 'A' turret, *King George V*. B. C. Cambray

Top left:
Cleaning ice off the 16in turrets of *Alabama*. This shot also gives a good impression of the great size of 16in guns. *US Navy*

Above left:
What it was all about — a convoy at sea. *IWM A3801*

Above:
***King George V* at anchor having been in collision with the destroyer *Punjabi*. The damage to the bows where the battleship slice through the destroyer is very evident; *KGV* had to go to Liverpool for repairs.** *IWM A9943*

Right:
US ships also assisted in covering Russian convoys. Here *Duke of York* and *Washington* form the covering force. *IWM A9297*

knew every blinking wave in that sea and his *exact* position was so-and-so, *not* the one he had been given! A double check showed that he was correct!' *B. C. Cambray*

The covering force usually operated quite some distance from the convoy it was protecting. The battleships had no part to play in protecting the merchant ships from U-boats and could themselves become vulnerable if subjected to air attack. Therefore, they left the convoy in the hands of the close escort of destroyers and cruisers, but remained poised to move in if there were any indications of a threat from German ships. Consequently, the covering forces largely escaped attack by the Germans but, from time to time, accidents occurred:
'During one operation in northern waters we did have one very unfortunate incident at sea which caused loss of life. We were out with our usual convoy covering force, together with American ships including the battleship USS *Washington*. *KGV* was leading in single line ahead with the destroyers stationed in one of their set-pattern close-screen formations. Visibility was poor, the Fleet was zig-zagging and, at the time, I believe was ordered to alter course by W/T.

'That day I had spent the forenoon on the flagdeck and was next on watch for the first dog and so happened to be on the messdeck just before early tea, when I felt the ship turn suddenly and appear to heave out of the water. Shortly afterwards loud explosions were heard along the ship's side and I immediately thought of a torpedo attack. By this time everybody was awake and I had managed to get life jacket, knife and bars of nutty from my locker. Over the broadcast then came the announcement: "This is the Commander speaking — Hands to Mine Stations!". My station was amidships near the Catapult Deck and everyone around me made their orderly way up through the various damage control routes leading to the upper deck.

'Once up top I could see that the visibilit was very poor and, although there wa plenty of activity on the signal decks, th ship appeared to be proceeding as normal However, I then observed a destroyer on th

Above:
Howe **at sea covering a Russian convoy. The 'King George V' class played a major part in the protection of the Northern convoy routes as they could pose a significant threat to *Tirpitz*, should the latter come out.** *IWM A15435*

Right:
Something else the battleships of the Home Fleet found themselves doing from time to time was taking VIPs across the Atlantic. Here President Roosevelt and Winston Churchill converse on board *Prince of Wales* in Placentia Bay, Newfoundland. *A. V. Godding*

port screen signalling to us to the effect that she was going to the aid of a crippled destroyer astern. Somebody told me that we had collided with an American destroyer but, in actual fact, it was one of our own, the "Tribal" class destroyer *Punjabi*, which had been cut in half by the *KGV*. The destroyer had primed depth charges on the stern and, as they rolled off during the collision, had gone off, and this was the cause of the explosion alongside *KGV*. But what was far worse for the destroyer men who had been in the rear half of the *Punjabi*, the depth charges going off had reduced the chance of survival.

'As I was given to understand at the time, the accident happened because the *Punjabi* was out of position at the end of the manoeuvre, probably not having carried out the correct zig-zag procedure when starting or completing the ordered course alteration. As it was, the sad fact remained that the *Duke of York* took over our duties and we returned to Iceland whereby survivors were transferred to us and subsequently we sailed to Scapa Flow . . . We shortly afterwards proceeded to Liverpool for damage repairs.'
G. Whyte

Six
Mediterranean Operations

1 Convoy Defence

Italy entered the war in June 1940. At that time the Italian navy possessed six battleships, *Conte di Cavour*, *Giulio Cesare*, *Caio Duilio* and *Andrea Doria* of World War 1 vintage, and the modern *Littorio* and *Vittorio Veneto*. The British had two battleship forces in the Mediterranean at this time: *Warspite*, *Malaya* and *Royal Sovereign* were based in Alexandria, which was to remain the base for the Mediterranean Fleet. At Gibraltar a force known as 'Force H' was gathered, consisting originally of *Hood*, *Valiant* and *Resolution*, although the composition of this force changed frequently.

The battleship war in the Mediterranean bore some similarities to events in the North Atlantic but there were also significant differences. As in the Atlantic, many of the operations were in support of convoy movements, especially as Malta became beleaguered and running convoys to the island became major operations. In addition, Mediterranean Fleet battleships carried out many shore bombardments long before the Home Fleet was significantly involved in such operations. Periods at sea in the Mediterranean were generally shorter than in the Atlantic, but whilst at sea ships were liable to attack at all times. Enemy airfields were never far away, especially when the Axis held most of North Africa. In addition, the Italian Battle Fleet was always likely to put to sea as a Fleet, something the Germans did not have the ships to do. Thus any Mediterranean operation, particularly a convoy movement, was always likely to lead to a Fleet action.

One of the first problems the C-in-C Mediterranean Fleet, Adm Cunningham, had to deal with was the problem of the French Fleet after the surrender of France. In the Western Mediterranean, Force H was obliged to shell the French ships lying in Oran, but in Alexandria the situation was not so difficult, although it was not clear at first what the French would do:

'The capital ships in Alexandria at that time consisted of *Warspite*, *Malaya*, *Royal Sovereign* and the aircraft carrier *Eagle*. Also in harbour were three French battleships [in fact one battleship *Lorraine* and four cruisers, plus some smaller ships], which presented a slight problem at the time of the French surrender. C-in-C Mediterranean never knew quite what they would do — try to fight their way out, blow themselves up or just give up peacefully. This is, in fact, what they did [give themselves up]. In case they

Below:
***Malaya*, one of the mainstays of the Mediterranean Fleet in 1940. She had not been as extensively modified as some of the other members of the 'Queen Elizabeth' class. In particular she had not been re-engined and re-boilered which, at 23.5kt, made her slower than some of the other battleships.** *IWM A7492*

Top:
***Renown* and *Ark Royal* operating together. It was in the Mediterranean that the battleships and carriers operated in the same force for long periods, and carrier air power was first used offensively.** *IWM A4015*

Above:
The Italian battleship *Littorio*. The three members of the 'Littorio' class were a significant threat to the British Mediterranean Fleet, and in particular British convoy operations in the Mediterranean. They mounted 9×15in guns and were capable of 30kt.
IWM45A

had tried to leave, we had our guns manned and some of us detailed off as boarding parties. During this time there were air raids, but no damage done to ships, but some damage to the town.' *T. W. Pope*

It was not long before the Mediterranean Fleet had its first brush with the Italians:
'The first week in July found the Fleet at sea covering a convoy bound for Alexandria. It was during this time that we met the Italian Fleet off the coast of Calabria, and action was joined during the forenoon and it lasted until late afternoon. Being below decks in the 6in battery, closed up on the guns, we could not see what was going on up top, but as the ranges were passed down and the order to load the 15in was passed, we knew that we were to engage the enemy.

'We [*Royal Sovereign*] were second ship in the line; ahead was *Warspite* leading, *Malaya* astern and *Eagle* well astern. I took a quick look through the layer's telescope and away out on the starboard side were our cruisers and destroyers, steaming at full speed with Battle Ensigns streaming out in the wind. It was quite a sight. Although *Warspite* and *Malaya* opened fire and the Italian shells were dropping close to us, we did not return fire, the reason being that our ship was too slow and could not keep up with *Warspite*. *Malaya* was ordered to take our place second in the line. One of the Italian ships was damaged and, under cover of a smoke screen, turned away and retreated. We followed for a time but the C-in-C concluded that they were leading us to a minefield, so discontinued the action.

'The Italian bombers had a go at us daily but again, although they dropped numerous bombs with some very near misses, no ship was hit. We had, through near misses, minor damage to the underwater part of the hull, mostly rivets missing, but sea water was getting into our fresh water tanks. It was because of this, and the fact that we were too slow to work with the fleet, that we left Alexandria and, with a stop in Aden, proceeded to Durban for dry docking.' *T. W. Pope*

The Italian opposition in this action off Calabria was *Giulio Cesare* and *Cavour*. It was fought at very long range, 26,000yd, and only the fire of *Warspite* was effective. She obtained a hit on *Cesare* which started a fire, after which the Italians retired.

In November 1940 Fleet Air Arm aircraft from *Illustrious* attacked the Italian Fleet in Taranto and succeeded in sinking *Cavour* and damaging *Littorio* and *Duilio*. *Cavour* was raised and repairs commenced but she did not see active service again. Both *Littorio* and *Duilio* were repaired and rejoined the Italian Fleet but, for a considerable period, the balance of capital ships was in the British favour.

In March 1941 *Warspite*, *Barham* and *Valiant*, plus the aircraft carrier *Formidable* were at sea protecting convoys en route to Greece. *Vittorio Veneto* with eight cruisers and 17 destroyers sailed to intercept the convoys, not knowing that the British were at sea in strength. *Vittorio Veneto* stumbled across some British cruisers and was subsequently attacked by torpedo bombers from *Formidable*, but without a hit being scored. The only hope for the British battleships to catch *Vittorio Veneto*, for they were a long way astern of the action, was for the Italian ship to be slowed down. A further attack by aircraft did achieve one hit on *Vittorio Veneto*, which slowed her down temporarily, although in the final outcome she was able to return to port without being caught by the British battleships.

However, the Italian cruiser *Pola* was hit by one torpedo from an aircraft and stopped in the water. Two other cruisers, *Zara* and *Fiume*, along with four destroyers, were sent back to her assistance on the evening of 28 March. J. G. Barlow was in *Valiant* at the time:
'At 22.10 that night the radar of *Valiant* picked up a stationary enemy vessel. The

Above:
Beneath a sky pockmarked with AA shellbursts, a Malta convoy is seen here under air attack with a 'Nelson' class battleship in the escort. *IWM A11201*

Left:
***Valiant* firing a salvo of 15in; the Battle of Matapan was fought at night when broadsides were fired — all the guns together rather than one gun in each turret at a time.** *IWM A23494*

C-in-C [Adm Cunningham in *Warspite*], receiving *Valiant's* radar report, turned the fleet to investigate. At 22.20 *Valiant* reported the bearing of the enemy ship at a distance of some 4½ miles. At 22.23 *Stuart* gave the alarm that other enemy ships were closing from a different bearing at about 2 miles distance.

'The enemy were units of the Italian Fleet, led by the destroyer *Alfieri*. Second in line was the 8in cruiser *Zara* followed by another 8in cruiser *Fiume* and the destroyers *Gioberti*, *Carducci* and *Oriani*. The cruiser *Pola* was the stationary enemy vessel which *Valiant* had picked up on her radar. (At this time it is said a visual sighting of the enemy was made by someone on *Warspite*.) [This is correct — the enemy ships could be seen from *Warspite's* bridge, including by the C-in-C himself.] At 22.27 the Battle Fleet opened fire: *Valiant* and *Warspite* fired broadsides at the Italian cruiser *Fiume* and sank her. *Valiant* then turned her guns on the cruiser *Zara* and hit her with the first broadside. *Warspite* and *Barham* then joined

Above:
***Malaya* being attacked with bombs in the Mediterranean during 1940.** *T. W. Pope*

the attack on *Zara* and brought concentrated fire on to the target, firing some 62 15in shells at the *Zara*. The badly crippled *Zara* was later torpedoed by the destroyer *Jervis*. By 22.31 (some 21 minutes from the first radar report) the Battle Fleet had either sunk or reduced the Italian force to burning wrecks.'

Roy Emmington was a boy in X turret in *Warspite* that night:
'I was in X turret's control cabinet. That was a night to remember because at night actions the 15in turrets fired broadsides. Salvoes shook the insides out of you so we were wondering what broadsides would do. [Salvoes were firing the two guns in each turret alternately; broadsides meant that both guns in a turret fired simultaneously.] We didn't have long to wonder: we got the orders to load both guns and we heard the breech workers shout "Ready!", and a few seconds after they fired. The feelings in a 15in turret firing broadsides would be hard to explain, especially the feelings of a seventeen-year-old boy. A number of broadsides were fired that night and two Italian cruisers, *Zara* and *Fiume* were sunk. It took us hours on the boys' messdeck to unwind that night.'

Thus ended the Battle of Matapan.

I remember one harbour we called at in the Med. The heads cleaner was a tubby man; on one night when the sailors were returning to the ship, the Commander was standing on the quarterdeck. Along came a rating pushing a wheelbarrow with old Tubby flaked out, his legs dangling over the front and his arms hanging over the back. His mate was staggering along and I can't remember seeing a Commander laugh so much.

The threat of air attack was ever-present whenever the Mediterranean Fleet was at sea. For those who manned the twin 4in mountings which were rather exposed, they faced not only enemy aircraft but also, on occasions, the blast of their own 15in guns:
'What was it like to be in action? Well I think, looking back, we were all as scared as hell but no Jerry was going to get the better of us, and so we quashed our feelings and just got on with the job in hand. I would point out here that my Action Station was on the iron deck as crew of the Mk 16 HA/LA anti-aircraft mounting. We had four: two to port, two to starboard. All were twin 4in guns with 16 men as crew to each mounting.

'Waiting for the firing order was always the difficult time, especially if you could see the aircraft approaching, but once in action one was too busy to worry, just keep the gun firing. For us we never had a misfire in action, but sometimes in training we did.

'Ready-use ammunition was to hand, but supply men had to worry about the empty shells from the guns dropping on their toes. The area around the gun was plastered with empty cases, but we managed always to keep the guns supplied.

'When the 15in guns went off, I and the others on the iron deck had to keep our rubber ear plugs in. We were exposed on the iron deck and several of us had a job to hear properly for a couple of days. The blast of the 15in was terrific: loud claps of thunder, brilliant flashes of orange and red flame, followed by black and grey smoke, and the ship would shake and shudder a great deal.' *R. V. Racey*

Most of the ships in the Mediterranean suffered damage from air attacks at one time or another. In May 1941 *Warspite* was badly damaged by a bomb during operations off Crete, to the extent that she eventually had to be withdrawn for repairs in the USA. Despite the extensive damage and the loss of 38 lives, there were some more light-hearted moments, such as when the Commander inspected the damage on the messdecks:
'Cdr Sir Charles Madden [who retired as an Admiral] came down to survey the bomb damage on the messdecks at Crete; He played up a bit, complaining about dhobeying all over the messdeck. His last standing order, "No dhobeying to be left to soak in buckets on messdecks".' *I. R. Nicholls*

Not only were the ships likely to be attacked by aircraft, but U-boats were also active. In September 1941 *Nelson* was hit by a torpedo:
'. . . Malta convoys and the torpedoing on the port side just forward of the armour plating. I remember her "whipping" as she was hit, and then carrying on as if nothing had happened, though somewhat down by the bows.' *J. V. Haddock*
[*Nelson* was sufficiently damaged to require repairs back in Rosyth.]

Barham was not so lucky: On 24 November 1941 she was hit by three torpedoes fired from *U331*. She very quickly listed heavily to

port, and then her after magazine exploded, destroying the ship. About one-third of her ship's company survived.

A further disaster occurred in December when *Valiant* and *Queen Elizabeth* were attacked by Italian 'Chariots' in Alexandria: 'About 06.00 I was woken abruptly by the pipe "Clear Lower Deck", so I hastily threw my blankets aside and, still in my pyjamas, ran up the companionway on to the quarterdeck. No sooner had I arrived on deck than I was given a stern "bottle" by a somewhat choleric Lieutenant-Commander who tore me off a strip for my stupidity in not putting on a dressing gown (it was quite a cold morning admittedly), and told me to go below at once and get it. It turned and dashed down the companionway again and was just putting on my dressing gown when I felt the whole ship shake from a very heavy explosion. In fact the deck where I was standing whipped up and down so violently several times that I had difficulty keeping my feet.

'I made my way up on to the quarterdeck again as quickly as I could, and as I did so I could feel the ship listing to port. The list increased to (I suppose) about 10 degrees but then stopped and, at the same time, the ship went down slightly at the bows, though not enough to make movement along the deck difficult.

'By this time the "buzz" had reach me that we had been attacked by an Italian midget submarine, but that the crew had been captured and were onboard. All eyes now turned on *QE* as she was believed to be the next target and a few seconds later I felt, rather than heard, a heavy underwater explosion from the direction of *Queen Elizabeth*, and a great cloud of black smoke poured out of her funnel. The explosion seemed to lift her bodily at least a foot or two, and when she settled back she was noticeably lower in the water but still on an even keel. She probably settled on the bottom as the water at Gabbari must have been pretty shallow.

'*Valiant* was moved later that day to the floating dock and stayed there under repair for about three months before eventually limping to Port Said, escaping through the Suez Canal and down to Durban for more extensive repairs.' *G. David*

Queen Elizabeth also required a prolonged period under repair. It was a bad day for the Mediterranean Fleet.

Above:
A battleship near-missed by a stick of bombs during a Malta convoy. *D. K. Bean*

Below:
***Barham* listing heavily after being hit by three torpedoes. As she capsized, her after magazine exploded, destroying the ship.** *IWM A46413*

As the Allies slowly gained the upper hand in the Mediterranean, the battleships were used extensively for shore bombardment. During the invasion of Sicily, *Howe* took part in one bombardment which provided some uncomfortable moments:

'The invasion of Sicily took place. My memories of that were being in Y magazine, loading cordite like the clappers as *Howe* shelled Trapani as a diversion for the main invasion force on the other side. I think our hearts missed a couple of beats as the Captain announced (he always told us what was going on up top) that for the next 20 minutes we would be going through a mined area. Sitting in a magazine just above the screws with the ventilation off was not the best of places to be just then.' *G. A. Langridge*

Later in the Italian campaign, *Warspite* was back in action after her repairs in the USA:

'En route we were ordered to the Salerno beach-head to bombard some 10-12 miles inland where Panzer troops were resisting the advance of the US forces with venom and tenacity. We immediately went to Action Stations and, during the night after altering course towards western Italy, we suffered heavy air attacks by waves of determined torpedo bombers (German). A running commentary of the actions was given over the Tannoy and, despite the continuous thunder of anti-aircraft pom-poms and 4in guns, it kept the crew well informed. From what I can recall, two or three torpedoes ran alongside, and only by skilful avoiding action we were not hit. The adrenalin ran high and the worst thing was that in the remote control office, we were battened down and could not see anything. However, three-quarters of the ship's company were in the same position. Action rations were issued at meal times — boiled eggs, sandwiches and kettles of tea. On arrival in Salerno Bay, small craft, destroyers, minesweepers and the odd cruiser were milling around. We immediately came under heavy air attack by Junkers 88s, but with heavy AA fire and covering fire from other ships, we suffered no damage.

'We then anchored quite near to the beach and, with our heavy armament, began to bombard the area in front of the gap being held by the Germans. A Marine Captain and a Leading Telegraphist were sent ashore to overlook the position and direct fire. The bombardment was a complete success and continued until a high-flying plane was spotted some 3,000 to 4,000 feet up, out of effective AA-fire range. It dropped a 3,000lb bomb and it was discovered afterwards that it was radio controlled. It hit the ship just astern of the hangar.

'I was on the messdeck at the time, about 11.55am, and about three compartments forward of the hit. It seemed as if an express train travelling at high speed had hit the ship; one tremendous explosion which seemed to jar every nerve in one's body. My immediate reaction was one of intense shock, followed by the desire to do something to allay my alarm. The ship seemed as if it had been picked up by some devilish hand, then dropped. Dust was everywhere and most of the crew rushed to the upper deck. A second explosion occurred, followed by a more intense shuddering, which must have been the bomb exploding in the region of the boiler rooms. 12 members of the engine and boiler room staff were killed and about 30-40 people injured. The stokers' messdeck was cleared

Below left:
***Valiant, Warspite* and *Barham* bombarding Bardia on 3 January 1941. The ships of the Mediterranean Fleet carried out a large number of bombardments, particularly in support of the amphibious landings in Sicily, on the coast of Italy and in southern France.** *IWM HU36427*

Below:
***Warspite* was severely damaged by a guided bomb whilst supporting operations off Salerno; she was out of action for over six months.** *IWM A9713*

Above right:
The last battleship operation in the Mediterranean was carried out by *Ramillies* when she supported the landings on the south coast of France in August 1944. *IWM A25722*

and turned into a temporary sickbay for the wounded. Most of the injuries were shrapnel wounds and very bad burns.

'Steerage way was lost and we were taken in hand by two small US Navy tugs. They could not hold the ship and we were drifting towards minefields, so the cruiser *Cleopatra* took us in tow and managed to pull us away from the bay into open water. The ship had settled about 12 feet in the water and damage control parties of the engine room branch did their best to prevent further flooding from the damage. It appeared the bomb had gone through about four decks and exploded underneath two boiler rooms which, it was reported, had just about dropped out of the ship. Later inspection showed a hole through which a double-decker bus could easily have been driven. We drifted off Salerno all night, held by *Cleopatra* who was unable to make any headway. Early next day an ocean-going tug came from Malta and took the ship in hand.'
R. W. Scott

In fact the ship suffered one direct hit and one very near miss which caused the flooding of five out of the six boiler rooms; *Warspite* took in some 5,000 tons of water. She was taken back to UK for repairs via Malta and Gibraltar. This damage was the most extensive suffered by a British battleship in World War 2 without the ship sinking, and shows what punishment they could withstand.

Battleship operations in the Mediterranean dwindled as the war progressed. The last action was support of the Allied landings on the south coast of France by *Ramillies* in August 1944.

'one evening in Algiers we were on the upper deck listening to records over the ship's broadcasting system when we got our eye on a rating coming along the jetty, labouring under the burden of a large sack on his back. We recognised him as from our ship, an old three-badge AB. When he duly reached us, we asked "What the hell have you got in the sack, Stripey?" "Budgie seed," he replied, without batting an eyelid. My oppo and I just looked at each other and laughed — we thought he was going round the twist. "Budgie seed?" we remarked. "Yes," he said. "Can't get it in the UK and, if you can, it's £1 a pound." He certainly had the last laugh as when we arrived in the UK he got it ashore alright. Next time we saw him he was waving a fist full of the old-time fivers at us; he must have made quite a handsome profit.'

Seven
Home Fleet

1 Engineering

One department of a battleship we have not yet looked at in any detail is the Engineering Department. Theirs was not the glamour or glory of the gunnery world above decks, but without them the ship could not function. Peter Stokes, who was an Engineer Sub-Lt in *Howe*, takes us on a tour of the department:

'Plumbers are necessary evils, required only to stop making smoke and to keep the quadruple screws turning as slow, as fast or as long as Authority on the bridge decrees. (A passage through the Suez Canal provides a comprehensive test.)

'The underworld of the battleship is rarely penetrated by the upper deck fraternity, possibly because of the likelihood of getting lost. Who can say with confidence in which of the four engine rooms or four boiler rooms he is at the moment? Who indeed could even find the entrance to the so-called Harbour Machinery Room, a large rectangular space sandwiched between forward and aft boiler rooms and flanked by the forward engine rooms? Many of the ship's company are believed to be unaware even of its existence.

'I am in the Harbour Room now, right in the middle of the ship, and as far down as you can get without getting your feet wet. It is not a place to linger unless watchkeeping duties compel it. It may be compared with Clapham Junction in that it receives, knots and despatches a complexity of [pipe] lines from and to all the surrounding spaces. Some, for superheated steam, are heavily wrapped in once-white lagging and look like enormous broken legs in plaster casts. Others, less bulky, convey salt water, fresh water, bilge water, boiler feed water, luboil and hydraulic fluid to their various destinations. The Harbour Machinery Room may also be compared to a tropical jungle: to be entered with caution, treated with respect and vacated as soon as possible.

'Several wild animals are permanent residents of the Harbour Room jungle, including huge beasts called evaporators which voraciously suck up a 24-hour diet of salt water and surrender, if in the mood, a few tons daily of fresh water for despatch to the ship's tanks. At not infrequent intervals it is necessary to descale the dinosaurs, a labour of Hercules involving the removal, treatment and replacement of most of their internal organs, generally under an anaesthetic administered not to the patient but to the operating team, in the form of an extra generous issue of iced goffers.

'Engine Rooms, Boiler Rooms and Action Machinery Rooms (the last-named housing turbo- or diesel-generators and the hydraulic pumps serving the gun turrets) all have to be approached from middle deck level — there are no low-level short cuts. The plumber, therefore, is usually slim and active by reason of the number of times he

Below left:
The manoeuvring platform of 'X' engine room, *King George V*, showing the large handwheels which control the throttles for the main engines. On the right, the cruising turbine was used when only moderate speed was required; the main turbine was used for high speed and for manoeuvring. *IWM A1784*

Below:
The boiler room of an 'Iowa' class battleship. On the left are the sprayers which spray the oil fuel into the boiler. *US Navy*

must climb near-vertical ladders in order to transfer himself from one space to another. Should this be a boiler room he must brace himself, for each high level entry has an air lock. He must open the first door and clip it shut; opening the second door will expose him to a blast of hot air at a pressure several inches above atmospheric and his ears will be assaulted by the noise of the two large forced-draught fans feeding air to the oil fuel burners at the boiler fronts. Keeping a watch of four hours in a boiler room in the tropics with a temperature of 125° is usually four hours too long, but that is a reward usually reserved at sea for a Chief Stoker, a Stoker Petty Officer and a squad of Stokers, their vigil only occasionally interrupted by the Second Engineer Officer of the Watch. ("Is everything all right, Chief?".)

'In comparison, the Engineer Officer of the Watch [EOOW] in the Controlling (X) Engine Room, providing he remembers to stand under the ventilation fan, has much more acceptable, even tolerable, surroundings. However, in the event of the 14in guns firing, it is prudent to move to one side as incoming debris may supplement the more usual dust and cordite fumes. When not seeking refuge in the telephone booth from the continuous high-pitched whine of the turbines, the EOOW may allow his gaze to roam over four sets of gauges informing him of the conditions in the other three engine rooms and boiler rooms. If he is lucky it will be a peaceful, even boring, watch, but at Action Stations it is advisable to remember that he and his staff are battened down in the citadel under six inches of armour plate, with all the hatches secured.

'In harbour, the Engineer Officer of the Day has unparalleled scope for involvement in chaos — a turbo-generator will trip (all lights start dimming), a "Red Alert" will require the flashing up of an extra boiler, the laundry machinery will "fall over", the firemain will develop a leak, the possibilities are endless. The day will culminate in Engineer's Rounds which, if carried out according to the book, will involve visits to every space manned by the Department, concluding with the writing up of the register. It is after such exertions that testing the Wardroom gin for purity, instead of the boiler feed water, will provide a brief relaxation before the Duty Picket Boat is towed in with a damaged propeller. Despite the hazards lurking in the Harbour Machinery Room, it's really more peaceful being at sea . . .'

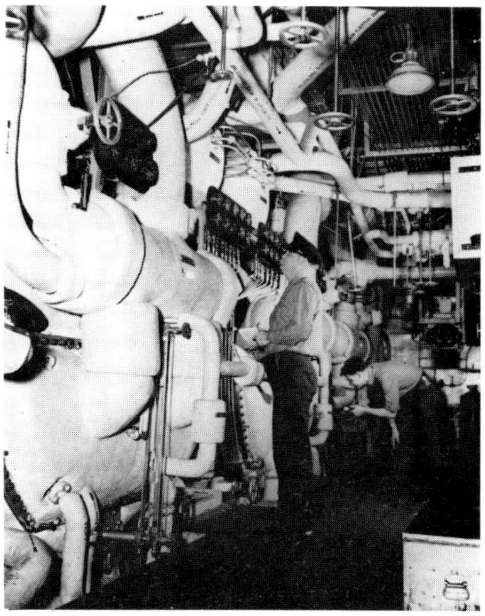

Above:
The evaporators in an American battleship, *Alabama*. Not only did the evaporators provide fresh water for the use of the crew by boiling sea water and condensing the steam, but they also provided feed water to top up the feed system for the boilers. Although the feed system is a closed system, ie the water recirculates after being condensed in the main engine condensers, there were always losses which had to be made up. *US Navy*

It was the stokers who had to do the bulk of the watchkeeping:

'My work was watchkeeping on the turbo-generators . . . I'm pleased to say that I was not alone in my ignorance of the new TGs. Once the teething troubles were overcome they more or less ran themselves. Also in the same machinery space was a set of hydraulics for the guns, and to make things nice and cosy one of the propeller shafts ran the length of the space . . .

'After a long stint on the TGs I changed part of ship to HPE [High Power Electrics] Party, in charge and responsible for the maintenance of about 80 motors and the lighting in various boiler rooms and machinery spaces.' *E. D. Sharpe*

'It was a must you did 8 months boiler cleaning, in three watches each day, when you first joined the ship. We started in A boiler room — by the time we finished F, we went back to A again.

'The locker flats, bathrooms and boiler room entries were on the next deck down to the messdecks. You were not allowed on the messdecks out of uniform. We came up from the boiler room, stripped off in the bath room, had a bath, washed your underwear in a bucket, changed into your uniform. The next watch you went down, you took the wet washing. The heat of the boiler room dried it out and it was ready to wear when you came off watch.' *G. T. Avery*

2 The Battle of the North Cape

The last action in which a British battleship engaged another battleship took place at Christmas 1943, in what became known as the Battle of the North Cape. In 1943 German capital ship operations were centred on northern Norway and the Russian convoys. However, by the end of the year *Lutzow* and *Scheer* had been withdrawn and were no longer a threat, *Gneisenau* was out of action in Kiel after damage by mines and bombs, and only *Scharnhorst* and *Tirpitz* remained in Norwegian waters. The latter had been badly damaged by a midget submarine attack in September 1943, although the British Admiralty could not be sure of her exact operational status. The threat from both ships remained, and the existence of that threat forced the British to allocate battleships to cover convoy movements.

Scharnhorst was a happy ship and occupied a position in German eyes akin to that which *Hood* had occupied in British eyes — she had a good reputation. In December 1943 she lay in Langfiord, around the corner from *Tirpitz*, where the men had to cope with the days and weeks spent at anchor. There were films, a library, card games, gambling, concert parties. An entertainments ship, *Emmanuel Rambur*, operated rather like the ships the British Grand Fleet had in Scapa in World War 1, coming alongside *Scharnhorst* from time to time. When the weather allowed, there was football and skiing, but in midwinter there were only two hours of twilight each day.

Scharnhorst sailed from Langfiord on the afternoon of Christmas Day to attack convoy JW55A which the Germans had detected southwest of Bear Island; five destroyers went with her. As close covering force with the convoy were the British cruisers *Belfast*, *Sheffield* and *Norfolk*. Covering Force 2, something over 100 miles

Right:
Adm Sir Bruce Fraser, C-in-C Home Fleet at the time of the Battle of North Cape. *IWM A16489*

to the southwest, consisted of *Duke of York* (C-in-C Home Fleet, Adm Sir Bruce Fraser), the cruiser *Jamaica* and four destroyers. The tactical problems facing the British were interesting, to say the least, in that *Scharnhorst* could outgun the cruisers and outrun *Duke of York*. If she came out, how to bring her to action?

But, as has been covered previously, the British had the experience of operating in the northern waters; they were hardened to the rigours. The German sailors were at sea for the first time for some months.

Intelligence that *Scharnhorst* might be at sea reached Adm Fraser in the early hours of the 26th. Although Fraser did not know her exact location, he could make fairly accurate judgements. *Scharnhorst* was, in fact, making for a position ahead of the convoy from which she could commence a sweep with destroyers to the southwest. A gale was blowing from the southwest with snow squalls and bitter cold, but the German ships were able to make 25kt to the north. They remained closed up at Defence Watches throughout the overnight passage; lookouts were alert. Searchlight control positions either side just abaft the armoured fire control position were used as lookout positions because they were fitted with very good optics, as were gun directors. The German Admiral, Adm Bey, did not expect too much from his 'Seetakt' radar. He also appeared to underestimate the potential of British radar which was to play a key part in the action. In addition, Bey had no intelligence of the presence of *Duke of York* in the area.

Top:
***Duke of York's* 10×14in guns were more than a match for *Scharnhorst's* 9×11in. *Duke of York* also had the advantage of superior radar, but *Scharnhorst* had the edge over *Duke of York* for speed. The ship in the background is *Glasgow* which was not present at North Cape.**
IWM A18000

Above:
***Belfast* who, along with *Norfolk*, firstly prevented *Scharnhorst* from attacking the convoy and then shadowed the German ship throughout the run to the south, passing locating reports which allowed *Duke of York* to intercept.** *Real Photos*

Right:
***Scharnhorst* heading north in Arctic conditions.** *IWM HU2219*

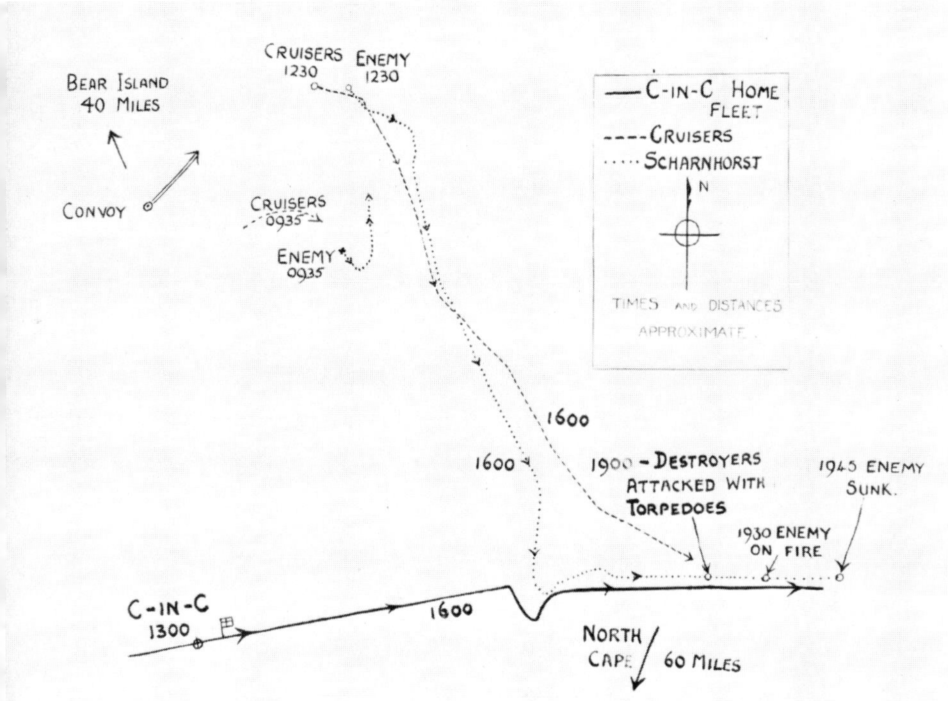

Above left:
The Battle of North Cape. *IWM A21091*

Above:
Rear-Adm Bey who commanded the German force consisting of *Scharnhorst* and her destroyer escort; the destroyers lost contact early in events and played no part in the final battle. Adm Bey was lost when *Scharnhorst* was sunk. *IWM HU46129*

Below:
***Jamaica*, operated with *Duke of York* throughout the battle.** *Real Photos*

At 07.30 on 26 December, Bey detached his destroyers to sweep to the southwest for the convoy. They lost touch with *Scharnhorst* and played no further part in the day's events.

Shortly after 09.00 on 26 December *Belfast* gained radar contact on *Scharnhorst* to the southeast of the convoy. *Belfast* and *Norfolk* closed and *Norfolk* opened fire, completely surprising *Scharnhorst*. Adm Bey ordered a turn away from the cruisers to the south, but correctly estimated that the convoy lay to the northwest. *Scharnhorst* therefore attempted to work round the cruisers in order to close the convoy again. However, *Norfolk* had obtained two hits in the first engagement, one of which carried away the forward 'Seetakt' radar, making *Scharnhorst* blind forward. Adm Burnett, in command of Force 1 (the cruisers), decided not to attempt to shadow *Scharnhorst*, but closed the convoy to cover against any subsequent moves by the German ship. Meanwhile, enemy reports were being passed to Adm Fraser in *Duke of York*, and it was an important feature of this battle that communications between the various British ships worked well.

Early in the afternoon *Scharnhorst* again closed the convoy, this time from the northeast, and was again detected on radar by the British cruisers. In another brief skirmish *Scharnhorst* was hit again, but *Norfolk* in return was hit by two 11in shells, and *Sheffield* suffered some damage from splinters. Adm Bey broke off and decided to terminate the operation against the convoy and return to Norwegian waters, then only some 9hr steaming away. As the news was passed amongst *Scharnhorst's* men, an air of relaxation descended onboard.

However, Force 1 followed the German ship on this second occasion and started shadowing by radar at long range, passing frequent reports by W/T to *Duke of York*. In fact, the shadowing operation by Burnett's cruisers was a classic piece of cruiser work. Meanwhile, in *Duke of York* the Admiral's staff had a clear picture of the situation and appreciated that they should be able to intercept *Scharnhorst* before she reached the Norwegian coast. On the Admiral's Bridge was the Admiral himself, the Flag Lieutenant, Fleet Torpedo Officer and Fleet Signals Officer. The latter had the important job of ensuring that the W/T links between the various British ships functioned smoothly. The Plot was immediately behind the

Above:
***Tirpitz* remained a threat to the northern convoy routes until finally knocked out by RAF bombing attacks in the fiords of Norway in November 1944. Here she is seen flying off a spotter plane. Her two forward turrets ('Anton' and 'Bruno') with their 15in guns can be clearly seen, as can the gunnery radar on the director tower at the top of the bridge structure.**
IWM HU1067

Admiral's Bridge, the forerunner of the modern Operations Room; this was controlled by the Chief of Staff. All information coming in from the Bridge Wireless Office, which was immediately adjacent, was plotted on the plotting table giving the Chief of Staff a clear picture of the situation. Also in the Plot was a new PPI (Plan Position Indicator) display for *Duke of York's* centimetric radar.

During the vital shadowing operation by Force 1, *Norfolk* and *Sheffield* both had to drop back, leaving only *Belfast* in touch with *Scharnhorst*. This put *Belfast* in a vulnerable position, because not only was it essential that she did not lose the German ship but also, should *Scharnhorst* turn on her, *Belfast* would be heavily outgunned. However, Adm Bey held on for the safety of the Norwegian fiords, whilst Force 2 headed her off. With both British forces were flotillas of destroyers who would play an essential part.

At 16.17 *Duke of York* obtained radar contact on *Scharnhorst* to the north-northeast, and then Fraser knew he could bring the German ship to action. He chose to hold on to close the range before opening fire, the initial detection having been made at about 22 miles; Fraser definitely had the tactical advantage at this point. At 16.50 *Belfast* was ordered to illuminate *Scharnhorst* with star shells. However, *Scharnhorst* could not be seen from *Duke of York*. The latter then fired her own star shell which burst precisely beyond *Scharnhorst*, silhouetting her against the pale light, her guns trained fore-and-aft, completely caught by surprise.

Duke of York immediately opened fire with her 14in and obtained a hit with her first broadside on *Scharnhorst's* 'Anton' turret. *Scharnhorst* replied within five minutes, using only 'Bruno' and 'Ceasar' turrets, and without the benefit of radar to find her target; she could only fire at gun flashes in the dark. Nevertheless, her gunnery was accurate.

Adm Bey quickly appreciated the nature of the threat which had appeared out of the dark to the south of him. But he knew he still had one tactical advantage — *Scharnhorst's* speed. Whatever his adversary was, he could outrun any British capital ship, so he turned to the east and did, indeed, start to open the range.

Fraser saw that the range was slowly opening and realised that *Scharnhorst* might slip from his grasp. There was little he could do unless the German ship could be slowed down. He ordered his destroyers to close but they had no real speed advantage over *Scharnhorst*. However, the opening of the range had one advantage — initially *Duke of York's* fire had a rather flat trajectory but as the range opened, the angle at which the shells fell increased and the plunging fire became more effective. A hit was obtained on *Scharnhorst's* No 1 boiler room and her speed reduced, temporarily to 8kt.

The action was to last 3½ hours, during which time *Scharnhorst* was hit many times, as *Duke of York* was now able to maintai range after the hit on *Scharnhorst's* No boiler room. The German seamen fough their ship hard, not only in gunnery term but also in terms of the efforts they made t overcome the effects of damage. They had system whereby the crews of disabled gun went to a personnel assembly centre fron which they could be sent to join other teams be they other gun crews, damage contro parties, etc.

The record of exactly what damag *Scharnhorst* incurred and when is, naturally unclear, but the hits from *Duke of Yor* appear to have caused:

- loss of No 1 boiler room, as alread mentioned. After the initial reduction speed was increased again to 22kt, but tha was not enough;
- loss of a 5.9in magazine;
- loss of all control position periscopes (The armoured control position was fron where *Scharnhorst's* Captain fought th ship. It had narrow viewing slits, shut b armoured shutters, and periscopes.);
- loss of No 3 engine;
- loss of at least one 5.9in turret;
- hot splinters penetrated 'Bruno' maga zine, which necessitated the magazine bein flooded. It was subsequently pumped ou again.

Finally, the destroyers were able to clos *Scharnhorst* and attack with torpedoes; th they did from very close range. It estimated that up to 55 torpedoes may hav been fired that day, with possibly 11 hits *Scharnhorst* started to list and slowed righ down; no ship could withstand that numbe of hits and *Duke of York* was still puttin

14in shells into her, fuelling the fires which raged in her superstructure. She finally sank at 19.45. 38 men were saved.

Tirpitz, the last of the German battleships in northern waters, was the target for numerous bombing attacks during 1944, during which she was heavily damaged. She finally capsized at her last anchorage near Tromso after an attack by Lancasters of RAF Bomber Command in November 1944 with 12,000lb bombs. However, the threat from *Tirpitz* was never great after the loss of *Scharnhorst* and, thus, the Battle of the North Cape was a significant moment in the naval war in support of convoy operations.

3 The Bridge

We have seen little of life on the bridge so far. This was the domain of Admirals, Captains, Officers of the Watch . . . and Signalmen:

'The bridge and compass platform of *Revenge* were wide open to the elements, as there was no cover whatsoever. There was little room to spare on the compass platform when the Captain, Officer of the Day, Officer of the Watch, Chief Yeomen, Yeoman, etc, were there, but there was plenty of room on the bridge at the rear of the compass platform. For the Signalmen, there would be a 10in signal lantern on either side, plus the Morse keys for the yardarm signal lamps.

'Two flights down from the bridge was the searchlight platform, and these 24in lamps were also used for long-range visual signalling. Flag signalling was conducted on the flagdeck, which was situated just forward of Y turret and under the main mast. At sea, flag signalling would be controlled from the bridge by voicepipe; whilst in harbour all signalling would take place on the flagdeck.

'Beneath the flagdeck was situated the Signal Distributing Office (SDO), often referred to as the nerve centre of the ship. In this office would be found direct telephone lines to various parts of the ship, and in harbour the shore telephone line was always connected in the SDO. The signal staff consisted of: 1 Chief Yeoman, 4 Yeomen of Signals, 3 Leading Signalmen and 12 to 16 junior signal ratings.' *E. Taylor*

Flagdeck work could mean being exposed to the elements for long periods:

'The time at sea was mostly spent on the flagdeck, with a Yeoman in charge and, apart from keeping a lookout, one of the main tasks was to man the signal halyards which, as the Admiral's ship, would be in constant use during daylight hours — which could be very long in the Arctic. If one let go a halyard, it was the custom for the unfortunate rating concerned to climb up the mast to retrieve it, weather permitting.' *G. Whyte*

Another facet of the signals organisation was the decoding of messages received by radio. G. A. Langridge, who was a stores rating by trade, found this one of his duties:

'When not at Action Stations, I did a cypher watch with an officer under usual watchkeeping routines . . . There was a marvellous American machine on which reels were fitted in a code sequence. I then typed out the numbers and the officer read the message off a ticker-tape coming out of the side of the machine.'

Returning to the bridge, there was one other

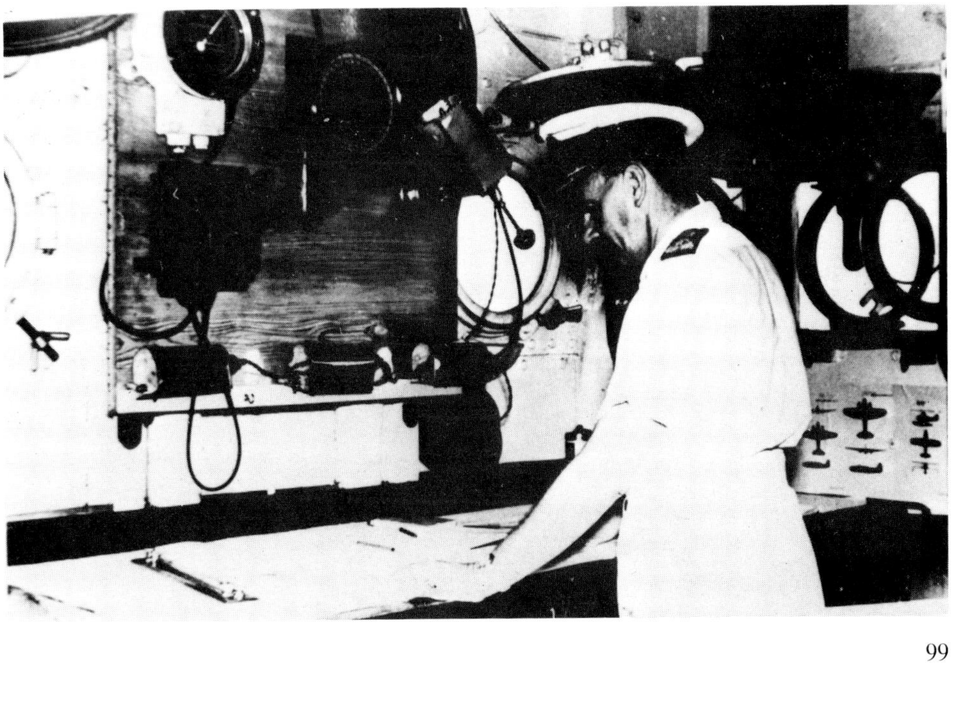

Above right:
The ship's bridge, in this case on board a 'King George V' class battleship. The magnetic compass is on the left, with the gyro repeat just left of centre. Either side of the gyro repeat are the voicepipes and also a steering telegraph, probably for emergency steering only. Ahead of the gyro repeat are the rudder indicator and engine telegraph repeats. Note the recognition silhouettes on the deckhead. *IWM A18115*

Right:
The Admiral's Bridge: this was separate from the ship's bridge; the ship could not be conned from the Admiral's Bridge — it was the position from which the Admiral could control the whole force under his command, with the appropriate plots and communications links readily to hand. This photograph shows Adm Sommerville on his bridge. Note, again, the recognition silhouettes, this time of aircraft. *A. Daniels*

person who knew it well — the bridge sweeper:

'One of my duties was as Admiral's Bridge sweeper. This was the élite of jobs. Only the Admiral, the Captain, the Commander, the Petty Officer in charge of the Admiral's Bridge and myself were allowed on the bridge. I cleaned mainly with brooms and cloths and my pet wirescrubber for the decks. I can see the wire scrubber in my hand now! That deck was white. I used to have great delight in stopping all Lieutenant-Commanders and below coming on to the Admiral's Bridge, with the specific orders entrusted to me. They climbed all the way up, only to go all the way down again!' *J. V. Haddock*

[The Admiral's Bridge was separate from the Ship's Bridge.]

But the Ship's Bridge was principally the domain of the Captain and the Officer of the Watch. John Haddock was later an officer in *Anson*:

'I can remember on that lovely bridge how I got to know where each telephone was by heart, and where it led to and from. But it was the calmness under all conditions that those senior RN officers instilled which I remember best. We had a brilliant Navigator, who was then Lieutenant-Commander Tibbits. On leaving Devonport after our refit — although we had tugs standing by — he never used them in taking that fine "KGV" class battleship out into the Sound. In a near-typhoon in the China Sea, he put *Anson* right alongside a Chinese junk which was sinking and all the occupants had to do was climb up a Jacob's ladder.

'One night during the middle watch (why is it that things always go wrong in the middle watch?) I was on watch together with a senior watchkeeping officer as we were in company with *Duke of York* on passage to Australia in the Indian Ocean. *Anson* was junior ship with all our destroyer escort screen out, forward, abreast and astern. We were on our particular pattern of zig-zag. As I "had the ship" at the time and followed these alterations — checking them off as we executed them after so many minutes — to my horror after we had altered course on one of these zigs, and so had the escort — *Duke of York* had not! This meant that if she held her course, we and *Duke of York* were going to meet!

'Having consulted with my senior OOW, telling him that the Captain and Navigator must be called, as per standing night orders, I pressed the necessary buzzers. Both the Captain and the Navigator were up on the bridge in next to no time. You can imagine how I felt! My natural reaction was had I made a terrible mistake, though I wondered whether the destroyers were possibly on another type of zig and all sorts of things had changed of which I had not been made aware. This is where sweat literally trickled down inside my vest on my back. Fortunately, as the ships were spread far apart, I had time to go through all my alterations with the Captain and Lt-Cdr Tibbits, and they, whilst watching *Duke of York* decreasing her distance between us, actually told me to hold my present course! You can well imagine my thoughts that should it be me who had made this terrible blunder with over 1,400 officers and men sleeping below . . .

'And then *Duke of York* altered course for the zig we were all on, followed by a signal giving profuse apologies for the mistake she had made. Capt Madden and Lt-Cdr Tibbits were so calm throughout this incident and gave "little me" such confidence in them, Capt Madden watching *Duke of York* with his monkey jacket on over his pyjama tops and wearing his pyjama trousers.'

The above incident, although slightly out of context as it happened in the Indian Ocean, was typical in that experienced RN senior officers found RNR and RNVR officers as their watchkeepers and in key positions in their ships.

4 D-Day

The final major action in which battleships of the Home Fleet participated was the D-Day landings, and subsequent support of the land operations in Normandy. *Warspite*, *Malaya*, *Ramillies*, *Nelson* and *Rodney* were all involved at various times. Battleship gunfire was not particularly effective against point targets such as gun emplacements, but was most effective against troop concentrations, formations of tanks, stores dumps, etc. With the aid of aircraft or ground observer spotting, long-range gunfire from battleships earned a reputation amongst the Allied soldiers for accuracy. Targets were engaged at extreme range on occasions, *Rodney* once firing at 17 miles. The amounts of ammunition used were enormous and the wear on gun barrels necessitated some ships having to replace them. *Ramillies* was off the beaches on D-Day itself. Although they did not know why, they found they had been practising for the operation for some time:

'Whenever possible, 15in firing practice was carried out with a target at or near the top of a hill. We had no idea why but, in my experience, sailors were more concerned with the happenings of the day, what's for dinner, etc, than the progress of the war.

'When we sailed for what was to be called D-Day, our Captain told us of the imminent invasion and that *Ramillies* was to silence German shore batteries on a hilltop. Only then did most of us put two and two together and, surprise, come up with four with regard to our frequent gunnery practice.

'The landings were postponed due to bad weather and we rolled around waiting. In a battleship we did not suffer as the small ships did, but we were in a state of excitement over what was to come. Our Captain was a believer in keeping his crew informed. He told us of our task and the opposition we could expect. He also told us he had received orders not to fire on any aircraft because THEY WOULD NOT BE ENEMY. I write that phrase in capitals because the pride, excitement, even joy in his voice, is a clear memory.

'When the great day came, *Ramillies* was creeping gently to her appointed position when she was challenged by light from a ship coming from shorewards. I should tell you that I had a grandstand view of all happenings, having just been "promoted" from layer of a 4in HA/LA gun, to director layer of a close range group. This director was in a sponson on the port side of the deck, immediately below the bridge, beside the Captain's sea cabin.

'Not receiving a satisfactory answer to her

Left:
***Ramillies* in action off the Normandy beaches on D-Day.** *IWM A24459*

challenge, the enemy ship opened fire. *Ramillies* returned fire in local control with all guns that could bear, the first one being the 4in I had just left. To see my gun banging away with somebody else pulling the trigger was annoying, to say the least. As soon as our 6in batteries trained round, they opened fire too and, our cover broken, the 15in opened fire. At this moment, presumably, the enemy vessel realised she had taken on a battleship and sensibly turned away, firing torpedoes as she went. Having no target to fire at, I was able to use my sights to watch all this and, on taking my eyes from the sights, saw the torpedo tracks scrape very close to our ship's side and disappear into the gloom. We watched in the direction they had gone and, before long, saw an enormous explosion. It was still too dark to see who caught the "fish".

'*Ramillies* continued her bombardment and my director crew sat, still with no target, as the light grew and we were able to see masses of aircraft, all ours, passing overhead, and gliders and parachute troops make their landings. We watched various individual actions fought.

'I do remember an important topic of conversation that first morning was what we could get to eat and when. We remained carrying out our bombardment, sometimes coming under return fire, until our 15in gun barrels were worn below an acceptable level. We then returned to Portsmouth where new guns were fitted incredibly quickly, and then returned again to France.' *C. T. Piddington*

Above:
***Ramillies* and *Warspite* off Normandy with Hamilcar gliders being towed overhead by Halifax tugs.** *IWM A23924*

Below:
Another view of *Warspite* off the Normandy coast. The advantage of gunfire support from the ships was that, firstly, it was available before artillery was established ashore and, also, targets well inland could be engaged with great accuracy. *IWM A23915*

By the end of 1944, a number of the older battleships had been placed in Reserve or a training role; *Royal Sovereign* had been transferred to the Soviet Navy. For the remaining battleships, the focus of attention had shifted to the Far East and Pacific.

Eight
Far East

1 Technical Improvements

Before we consider battleship operation in the Far East and Pacific, it would be appropriate to review some of the technical improvements which were made to the ships as the war progressed, and which were to reach their zenith in operations against Japan.

On the gunnery side, the main changes made to armament during the war were substantial increases in anti-aircraft protection. The threat from aircraft had first become clear in the Mediterranean and was an even greater factor in the Far East. At every opportunity, additional pom-poms and 20mm Oerlikon mountings were fitted to ships. For example, by the end of the war, *Duke of York* had 88×2pdr pom-poms, 8×40mm Bofors and 55×20mm Oerlikons.

Radar was a significant development during the war. We have seen how it played a major part in the Battle of the North Cape, and the ships in the Far East had comprehensive radar fits for early warning for both aircraft and surface contacts, and also for gunnery direction and fire control. Communications also improved. Matapan had been the first occasion when battleships were manoeuvred in battle using short-range radio telephony, ie speech rather than Morse.

The ability to replenish at sea was fundamental to operations in the Far East; the skills and the equipment required improved, so that by 1945 ships could stay at sea for long periods, supported entirely by tankers and stores and ammunition ships, the 'Fleet Train'.

Finally, aircraft should be considered. All battleships carried aircraft at the beginning of the war and we have seen how the German raiders used their aircraft in

Below:
This photograph of *King George V* taken in Sydney shows the wide variety of armament fitted by the latter stages of the war. On the left is the secondary director for the main armament, with the rake-shaped aerial of its radar. To the right is a high-angle director, also fitted with radar. On the deck below, from left to right, are a single 20mm Oerlikon, two twin 40mm Bofors and an eight-barrelled pom-pom. On the upper deck are two of the 5.25in turrets. At this stage of the war, the anti-aircraft armament of the 'King George V' class was, typically, 88×2pdr pom-poms, 8×40mm Bofors and 55×20mm Oerlikons. *J. E. Bowen*

searching for targets, how *Warspite* used her Swordfish at Narvik, how *Malaya* spotted *Scharnhorst* with her aircraft. *King George V* originally carried the Walrus:
'This was a great occasion when it was launched. Everyone possible appeared on deck and Winnie shot across the catapult like a bat out of hell — and just about had sufficient speed to get up. On the odd times she was launched when at sea, she was brought back by the ship steaming pretty quickly and then turning, the "skidding" having the effect of creating a large, smooth area into which she could land. Then came the difficult task of hitching her to the crane, a dangerous and unpleasant task for the rating concerned.' *B. C. Cambray*

As the war developed, and particularly in the Far East, battleships operated more and more as members of aircraft carrier task forces, and therefore their own aircraft were surplus to requirements. In fact, by the end of the war all aircraft had been removed.

Above:
'Winnie shot across the catapult like a bat out of hell — and just had sufficient speed to get up.' A Walrus being launched from *Rodney*. By the end of the war, aircraft had been removed from battleships: the carriers could do the job instead. *IWM A16075*

Below:
Refuelling at sea was an essential skill for operations in the Far East; here a destroyer is being refuelled by a battleship. The battleship would herself fuel from a tanker whilst under way; the escorts were frequently refuelled from the battleships. *S. Pearson*

Unfortunately *Repulse* and *Prince of Wales* were not part of a carrier task group when they deployed to the Far East in December 1941, as the British response to the Japanese threat. Unprotected from air attack except by their own guns, they were to pay the price. The detailed account of events that day has been told many times but this book would not be complete without the stories of two men who survived the loss of *Repulse* and *Prince of Wales*:

'The early hours of 7th December, Action Stations were sounded. We thought "not another exercise", but it was the real thing. Jap planes raided Singapore, dropping a couple of bombs on the town, no doubt to sink morale. We left Singapore on the 8th December, with the *Prince of Wales* and four destroyers to intercept Jap convoys.

'On 10th December, I had the forenoon watch on the Port Condenser Room, looking after three sets of evaps [evaporators]. There were 7 of us on 08.00-12.00. We were at Action Stations, Red.

'The action started about 11am. Between the Condenser Rooms was a porthole. During the action, I saw my best mate Paddy through the porthole. He waved, then there was a load of smoke around him — his compartment had been hit. I knew he had gone. I shed a few tears. We didn't get relieved that watch. The ship was zig-zagging a lot and listing. Around 2pm I wasn't getting much steam on the evaps which meant boilers out of action or cold water getting in. I said to the Chief Mech we should leave the compartment as I thought the ship was sinking. He agreed after a while.

'We had to open a big armoured hatch to get out of the halfdeck. There was another hatch with a manhole cover open, leading to the quarterdeck; all we could see was the water rushing back into the messdecks. When I crawled out, the force of water pushed me against one of our torpedoes. I managed to grab the fin; the force of the water broke a ring on my finger. I found out later that only 2 of us out of the 7 got away; out of 82 below, only 9 of us survived. The ship was listed right over, the funnels in the water. We crawled up the deck and walked down the ship's side. One officer onboard told us to watch the screws as they were still going around. I jumped in and swam away. Our main worry was sharks: I passed three which must have been dead or stunned with the shock of explosions down below. I passed the raft with Capt Tennant on board and other ratings. There were too many for anyone else to get on. I was picked up by the *Electra* [destroyer] and was told I had swum about two miles. I saw the *Repulse* go down while I was swimming. She went down like a good ship should, stem up, stern down, then the bows last.' *G. T. Avery*

C. T. Piddington was also in *Repulse*, but was able to see more of what went on:

'I believe we all thought *Repulse* was unsinkable. We had impressive fire power, both surface and anti-aircraft. When in company with *Prince of Wales*, we were sunk. It was hard to believe it was happening.

'Much rubbish has been said and written about the loss of these two ships. As a lowly Ordinary Seaman at the time, I can't claim to know a great deal about decisions made

Above:
***Repulse* suffered from the problem all the battlecruisers had — insufficient armoured protection, particularly as her reconstruction in 1936-39 had only improved the deck armour. Her anti-aircraft armament was improved, but was weak by the standards of the latter years of the war. However, *Repulse* was lost to aircraft attack in 1941.** *IWM A445*

Below:
***Repulse* at speed; even in 1941, at 32kt she was still a fast ship.** *P. Bosworth*

by Admirals but our Captain, Capt Tennant, a great man and my first and lasting Naval Hero, always spoke to the ship's company and told us all he could about what was going on. He told us, in Singapore, reports had been received of an enemy convoy, supported by battleships, approaching the Malay peninsula. The Admiral had to decide whether to send his two capital ships, knowing air support could not be provided, or do nothing. The choice was obvious and we duly sailed. Before long we spotted enemy aircraft that stayed out of range but kept in sight. We all knew we could expect to be attacked. Capt Tennant spoke to us all again and spelled this out.

'When the attacks came, I was on watch in a 15in turret, "A" turret to be exact, and had quite a good view through the control window. The first high level wave hit us with a bomb on the catapult deck, starting a fire. *Repulse's* evasive steering prevented further hits until low level torpedo attacks started. There were no suicide attacks. The Japanese sank us through plain bravery. A line of planes would start to come in, it seemed quite slowly, being shot out of the sky but still kept coming. I knew we evaded several torpedoes but finally one struck us aft and we lost steering. After that, they came thick and fast. I tried to keep count and it seemed there were 19 such hits but, allowing for excitement, it was probably less.

'About this time the 15in was not being used, so a party, including myself, was sent below decks, down through the working space, to assist in AA ammunition supply. I cheerfully worked away, still confident of *Repulse's* invulnerability, until I was sent back to "A" turret in some disgrace, having admitted to a Petty Officer my anti-flash gear and lifebelt were still in my locker. Shortly after this, the ship began to list heavily and "All hands to the upper deck" was broadcast. On leaving the turret, we were met by a hail of machine gun fire which caused us to scatter, and I realised, for the first time in my life, it COULD happen to me. On looking around, it appeared to me that the sea was tipping up at a sharp angle and, if I jumped in, I'd slide down and disappear over the horizon. I'd not realised the ship had listed so far.

'Cutting a long story a bit shorter, I was eventually picked up by one of the escorting destroyers, *Electra*, where everyone was very kind to us. My sharpest memory of that event was standing naked, except for a thick coating of black oil, and thinking how wonderful everything was. There I was, almost unhurt, the sun was shining and my special friend (who I'd been sure had been killed) was standing beside me.'

3 Pearl Harbor

Three days before *Prince of Wales* and *Repulse* were sunk, the Americans had suffered a greater disaster at Pearl Harbor; the story of the Japanese attack on 7 December 1941 is well known. However, what is not so well known is the effect the attack had on the US Navy's battleship force in the Pacific. There were eight battleships in Pearl Harbor that day, the whole of the Pacific Fleet battleship force apart from *Colorado* who was refitting at the time. Of the eight, *Arizona* was sunk and never salvaged; she still rests in the shallow waters as a national monument. *Oklahoma* was also sunk but she was salvaged, although she did not become operational before the end of the war. *Nevada*, *California* and *West Virginia* were all victims of torpedoes and they too were sunk. They were all salvaged and repaired and did rejoin the fleet before the end of the war. In fact, *Nevada* was operational again in 1943. *Tennessee* and *Maryland* suffered moderate damage and were both repaired, the latter being ready by February 1942. Finally, *Pennsylvania* was in dry dock at the time of the attack and only had slight damage and was soon fully operational. However, for a period there were no US battleships operational in the Pacific.

At the start of World War 2, the US Navy possessed a total of 15 battleships. As we have seen, nine were in the Pacific, the other six were in the Atlantic; none of the ships had been completed later than 1923. In common with the 'Queen Elizabeth' and 'R' classes of the Royal Navy, some of the ships had been modernised in the 1930s. The 'New

Left:
Pearl Harbour: this photograph shows *Pennsylvania* in dock with the destroyers *Downes* (left) and *Cassin* (right). *Pennsylvania* was the only battleship to get away with only light damage and she was soon operational again. *IWM OEM3605*

Right:
Like the 'New Mexico' class, the two ships of the 'Nevada' class were built in World War 1 and modernised during the 1930s. *Nevada* (BB-36) was sunk at Pearl Harbour: she was salvaged and was operational in 1943, having been modified during the repair period. She served in many theatres including the Atlantic, off Normandy, off the south coast of France and in the Pacific. *Real Photos*

Below:
***Idaho*, a battleship of the 'New Mexico' class. This class was built during World War 1 but were extensively modernised between the wars; even so, *Idaho* was the only one with a modern anti-aircraft armament. She was in the Atlantic at the time of Pearl Harbour, but transferred to the Pacific in 1942. The main armament was 12×14in guns in triple turrets.**
Real Photos

Mexico' class (*New Mexico*, *Mississippi* and *Idaho*) had been modernised the most extensively. A limitation from which all these older ships suffered was the lack of speed, all having a maximum speed of about 21kt.

In the Atlantic, US ships assisted in the protection of the transatlantic and Russian convoys. They also played their part in supporting all the major landings, both in the Mediterranean and in northwest Europe. For example, *Arkansas*, *Texas* and *Nevada* all played a part in the Normandy operations. In these areas, the role of the ships and life on board was very similar to Royal Navy battleships. The convoy routines were the same and combined forces of American and British battleships supported Russian convoys; bombardment operations were exactly the same as the British ships. No American battleships suffered significant damage in Atlantic or Mediterranean operations.

The war in the Pacific had many features which made big differences to the employment of the battleships and the nature of life onboard. Firstly, the distances involved were much greater than in the Atlantic and ships operated far from their bases. Secondly, it was the aircraft carrier which was the major fighting ship in the Pacific campaign. The battleships were largely employed in supporting roles, of which more in due course. And finally, the Pacific campaign became one of 'island hopping', a series of amphibious operations which slowly chipped away at the outer defences of the Japanese conquest until, at last, the mainland of Japan itself came under attack. However, the early battles in the Pacific after Pearl Harbor, notably Coral Sea and Midway, did not involve any battleships: these were the first of the major carrier battles. It was to take some time for the battleship force to get into action.

4 British Far East Fleet

Whilst the US battleships were recovering from the effects of Pearl Harbor in the Pacific, the British Admiralty was attempting to fill the vacuum left in the Indian Ocean. All they could spare were three of the older ships: *Revenge*, *Royal Sovereign* and *Ramillies* were the first arrivals in the Eastern Fleet in early 1942, to be joined shortly thereafter by *Warspite* and *Valiant*. The force was in no state to take on a major Japanese attack in the Indian Ocean, particularly an attack by aircraft carriers. All that could be hoped for in the early days in the Indian Ocean was to ensure the safety of the bases in Ceylon and on the east coast of Africa, although even Ceylon was shown to be vulnerable when the Japanese carried out an attack in early April 1942. Fortunately they did not return.

For the ships' companies, the more immediate problems were caused by the difficulties of living in old ships in very hot conditions:
'I have some rather vivid memories of life on board in the tropics, especially whilst at sea. Messdecks were, at best, very overcrowded and, coupled with the continual steamy heat below decks, it's hardly surprising that tempers frequently rose to the point when fights broke out.' *A. Daniels*

Conditions in the tropics were bad and every available minute was spent on the upper deck. Hammocks were rarely used as such, but were laid on the deck in any cool place, the recreation space, the waists and the bridge superstructure. Lice became a problem: they managed to infiltrate some of the hammock nettings . . . The bathrooms were as good as saunas and water sometimes in short supply.' *R. W. Scott*

Lice were not the only pests:
'Ships, and not only battleships, had their own brands of wildlife. Cockroaches were the most numerous, and what was most puzzling was why they always lost their foothold on the deckhead when directly over your plate!

'The tropical heat was the ideal conditions for breeding weevils and many a female gave birth in the flour store. They multiplied very quickly and proceeded to infest the whole stock of flour. It was the only time in my life when I ate bread with a built-in meat filling.' *A. Daniels*

In early May 1942 the battleships of the Eastern Fleet supported landings on the island of Madagascar and the island was occupied by Allied troops. At the end of May, *Ramillies* was at anchor at Diego Suarez Bay. W. E. Broomfield was Coxswain of the Duty Picket Boat:
'30 May — I was Coxswain of the Duty

Right:
Tropical life: the white helmets did not last long and were discarded as headwear. This group is on the 4in gun deck of *Revenge* in the Indian Ocean. *E. R. Moyle*

Picket Boat — brass funnel job, bit of a relic — broke down too often. We had a lull about 20.20 then "Away Picket Boat's crew" was piped for a mail run. *Manxman* was 8 miles away with mail on board. It was a warm and clear night, with a huge tropical moon shining. *Ramillies* lay in a golden ribbon of moonlight — a veteran queen. "One-funnelled baa-lamb" we called her.

'Whilst alongside *Manxman*, we heard low-flying aircraft. When we got back to *Ramillies* anti-aircraft guns were manned. They stayed closed-up for about 2 hours, then fell out.

'31 May — I sat at the mess table answering letters when suddenly there was an ear-splitting crash, a tremendous shudder, a blinding flash and then darkness. There was shocked silence for about 3-4 seconds, then uproar. Groping hands found the guide rope overhead, placed there for just such an occasion. It led out of the messdeck. I reached the upper deck. Men were going to their Action Stations and Emergency Stations. Bugles were going and orders being shouted.

'*Ramillies* had been torpedoed. The Bosun's Mate piped "Away both Picket Boats' crews and Both Charge Parties". We had 2 depth charges per picket boat. Whilst we were patrolling, we spotted a ripple and a periscope; the Picket Boat went full ahead and ran over the periscope. The depth charge was dropped. No wreckage appeared but it may have been successful. It was a very long night.

'When I got back to the messdeck, lockers were overturned, drawers out, water covering the deck, mess tables collapsed and things floating in the water. There was only dull secondary lighting.

'*Ramillies* had sunk on to the bottom. The torpedo had gone through the 4in magazine, right through the 15in magazine and into the 15in shell room, but the only casualty was from shock. [The torpedo itself did not penetrate the above compartments, only the effect of its detonation, which was against the hull.]

'With wire cables wrapped round the ship to prevent the armour plating falling out, we managed to reach Durban and docked. The hole was 30ft×29ft — we de-ammunitioned through it.' [*Ramillies* had, in fact, been torpedoed by a Japanese midget submarine.]

Apart from the Madagascar operation, the Eastern Fleet was not in a fit state to undertake offensive operations, so:

'We swung round the buoys in Mombasa and Trincomalee sometimes for weeks on end. We also were alongside in Durban several times and we were in dry dock there nearly five months. We [*Revenge*] had a nickname in Durban — HMS *Wallflower* — but we did not mind one bit, it was a lovely place to be.' *R. V. Racey*

Durban offered the delights of a pleasant climate, hospitable people who did much to look after the sailors, and none of the restrictions of food rationing. Mombasa was rather different:

'The base at Mombasa was established in Kilindini Harbour. Swimming and an occasional drink at the huge canteen were the only recreations available. A meal at the Palm Court Hotel was an expensive treat.' *R. W. Scott*

. . . and at Trincomalee:

'Trincomalee was a huge bay surrounded by jungle. There was a NAAFI ashore and a stores area and a headquarters. We spent our time training and undertaking routine duties on the ship; there was fuelling and storing to be done. The main recreation was swimming.' *W. H. Smith*

In Trincomalee, an American seaman called to our messman, 'Say Limey, when did you leave the old country in this old spud boat?' The messman replied, in a droll voice, 'Yesterday, and, by hell, it was raining.'

Above:
AB R. Racey ashore in Durban. This port was a welcome haven from the war where the people were hospitable and the climate pleasant.
R. Racey

Right:
All four ships of the 'Royal Sovereign' class in silhouette off Madagascar. *E. R. Moyle*

5 US Operations

It would take the British time to build up significant strength in the Indian Ocean, but for the time being the defeat of Germany had priority. However, by the summer of 1942 American forces were in a position to carry out the first of many amphibious assaults, on Guadacanal in the Solomon Islands. The battle for Guadacanal stretched over several months, with considerable aircraft carrier activity. In addition, American and Japanese battleships were to clash for the first time off Guadacanal.

The Japanese had started the war with 10 battleships but, like the Americans, all their ships were at least 20 years old. Most of them had been modernised in the 1930s in the same way that the Americans had modernised their ships. However, the Japanese ships had one advantage over their US equivalents — they were all faster. Indeed, the oldest of the Japanese battleships, the 'Kongo' class, had been fitted with completely new propulsion systems in their modernisations and were capable of 30kt. This high speed meant they were capable of operating with the carrier fleet, a role in which they were employed.

Both the US and Japan had battleships building when war broke out. The two ships of the 'North Carolina' class, *North Carolina* herself and *Washington* became operational in 1942, along with three ships of the 'South Dakota' class (*South Dakota*, *Indiana* and *Massachusetts*) as did the first of the new Japanese ships, the giant *Yamato*.

On the night of 12/13 November 1942, the Japanese planned to bombard Guadacanal, followed by the movement of a large number of troops loaded in destroyers to the island. The bombardment force included the two 'Kongo' class battleships, *Hiei* and *Kirishima*. Early in the morning of 13 November, the Japanese ships clashed with a force of American cruisers and destroyers. The US ships were severely

Above:
The Japanese battleship *Kirishima:* four ships of the 'Kongo' class had been completed in 1913-15, but had been reconstructed in the 1930s. In particular they were re-engined and reboilered which gave them the speed to operate with the carriers. *Kirishima* had to be scuttled after being heavily damaged by *Washington* off Guadalcanal. Her main armament was 8×14in guns and she had a maximum speed of 30kt. *IWM MH5923*

battered but they managed to damage *Hiei* with torpedoes. This ship was subsequently sunk by aircraft after repeated attacks in daylight on the same day.

The Japanese continued with their attempt to reinforce Guadacanal during darkness on the following night. Raizo Tanaka, one of the Japanese commanders, describes the engagement:

'. . . late afternoon, dispatch from Commander-in-Chief Combined Fleet ordered that we continue directly toward Guadacanal. Unusually successful radio communications at this time provided information that Second Fleet was advancing at full speed to attack the reported enemy fleet. This meant that the fleet flagship *Atago* [cruiser], battleship *Kirishima*, two ships of Crudiv 4, and several destroyers would be supporting our effort. Thus it was with a feeling of relief that I gave the order to proceed with the operation. By sunset I was further heartened by the sight of several of my rescue destroyers, filled to capacity with army troops, catching up with my depleted force. Shortly after midnight, with visibility at seven kilometres, we were greatly encouraged to sight our Second Fleet main body dead ahead. With these stalwart guardians leading the way, we continued the advance.

'Approaching from east of Savo Island, our van destroyers were first to engage the enemy, opposing several heavy cruisers. Heavy gunfire ensued and the entire vicinity was kindled by flare bombs. We could see individual ships set afire — friend and foe alike. *Atago's* searchlights soon played on enemy vessels which we were surprised to find were not cruisers, but 'Washington' class battleships! This then was the first battleship night action of the war [in the Pacific]. *Atago*, *Takao* [heavy cruiser] and *Kirishima* loosed their guns in rapid succession and the enemy opened return fire. I chose this moment to order a northward withdrawal of the transports, feeling that for them to continue into the battle area would only add to the confusion. At the same time I called for three ships of Desdiv 15 . . . To advance and attack the enemy.

'An hour past midnight this battle, which had started and ended in darkness, was over. It is believed that the enemy had lost two heavy cruisers and one destroyer sunk, one heavy cruiser and one destroyer seriously damaged. When my ships reached Guadacanal a burning heavy cruiser of the enemy was observed. We were of the opinion that two enemy battleships were damaged by torpedoes . . . We suffered the loss of battleship *Kirishima* (her crew were rescued by destroyers) and destroyer *Ayanami*, but felt that this Third Battle of the Solomons had ended in our favour.'

In fact, Tanaka's force had come upon *South Dakota* and *Washington*. In the initial skirmish when the Japanese illuminated with searchlights, *South Dakota* was seriously damaged by gunfire, sustaining 42 hits. However, *Washington* opened very accurate fire, entirely on radar information, on *Kirishima* at a range of 8,400yd. In the space of a few minutes, the Japanese ship was hit by some nine 16in shells and about 40 5in. She was subsequently scuttled. *Washington* was unscathed.

Guadacanal did not finally fall into American hands until February 1943. The Americans then commenced their 'island hopping' campaign which was to take them up through the Solomon Islands to New Guinea in the south, and via the Gilbert and Marshall Islands, the Marianas and Saipan to Guam in the central Pacific. From these positions they were ready to launch an assault on the Philippines.

6 Indian Ocean I

Meanwhile, in the Indian Ocean at the beginning of 1944, the old 'R' class ships had been replaced by *Queen Elizabeth*, *Valiant* and *Renown*. At last it was possible to take the offensive:

'Early in 1944 destroyer reinforcements arrived to join the Eastern Fleet. In March 1944 the American carrier *Saratoga* arrived from the Pacific to join Adm Somerville's command. The first strike of this force supported by *Illustrious*, *Valiant* and *Queen Elizabeth* was on the island and port of Sabang in early April 1944. The airfields of nearby Sumatra were bombed and bombarded with 15in shells. A good deal of damage was done and the Japanese were attacked for long periods. They were unable to counterattack until late that day when a weak effort was easily warded off without casualties. The fleet returned to Trincomalee.' *W. Burley*

Some recollect this attack in rather more straightforward terms:

'We would operate from Trincomalee in what were known as "Club Runs". We would go down with one or two carriers and some cruisers, and we would go and attack Japanese bases in Sumatra in particular . . . We really started to learn our job against the Japanese for our eventual part in the British Pacific Fleet.' *D. Higgins*

The Eastern Fleet did indeed carry out another 'Club Run' shortly thereafter:

'In less than a month after this, the fleet was again on an offensive strike, this time to Surabaya over 2,000 miles away. Aircraft from *Illustrious* and *Saratoga* attacked and damaged the airfields in the early hours. They also succeeded in damaging enemy aircraft and, later in the day, severely damaged oil refineries. Battleships *Valiant*, *Queen Elizabeth*, *Richelieu* and the battle-cruiser *Renown* bombarded and left, leaving aircraft destroyed and ships sunk, without any loss to the fleet. These Allied attacks continued at about one per month during the remainder of 1944, and now included the Anderman and Nicobar Islands.' *W. Burley*

Right:
Unicorn, Queen Elizabeth, Illustrious and Valiant as seen from Renown in the Red Sea, when the Indian Ocean forces were being strengthened with the arrival of more modern ships to replace the 'Royal Sovereigns'. *A. Daniels*

Below:
The silhouettes of *Valiant* and the French battleship *Richelieu* as seen from the quarterdeck of *Queen Elizabeth*. *Richelieu* had gone over to the Free French after lying in Casablanca since 1940. She was refitted in New York in 1943 and then joined the British fleet in the Indian Ocean; her 8×15in guns were a welcome addition. *A. Daniels*

7 Leyte Gulf

The Americans launched their assault on the Philippines at Leyte Gulf on 20 October 1944. It is outside the scope of this book, however, to consider all the complex naval operations and actions which took place after the strong Japanese response to the American invasion. We must be limited to considering the parts played by the battleships on both sides, 21 of which took part in the Battle of Leyte Gulf.

Japanese forces were divided into four groups: however, only three of the groups included battleships. The first Japanese

Right:
A ship of the 'Ise' class under aircraft attack off the Philippines. This class can be recognised by the built-up stern which was, in fact, a flightdeck designed to take 'Judy' dive-bombers, but there were never enough aircraft or pilots for them. *IWM NYP46113*

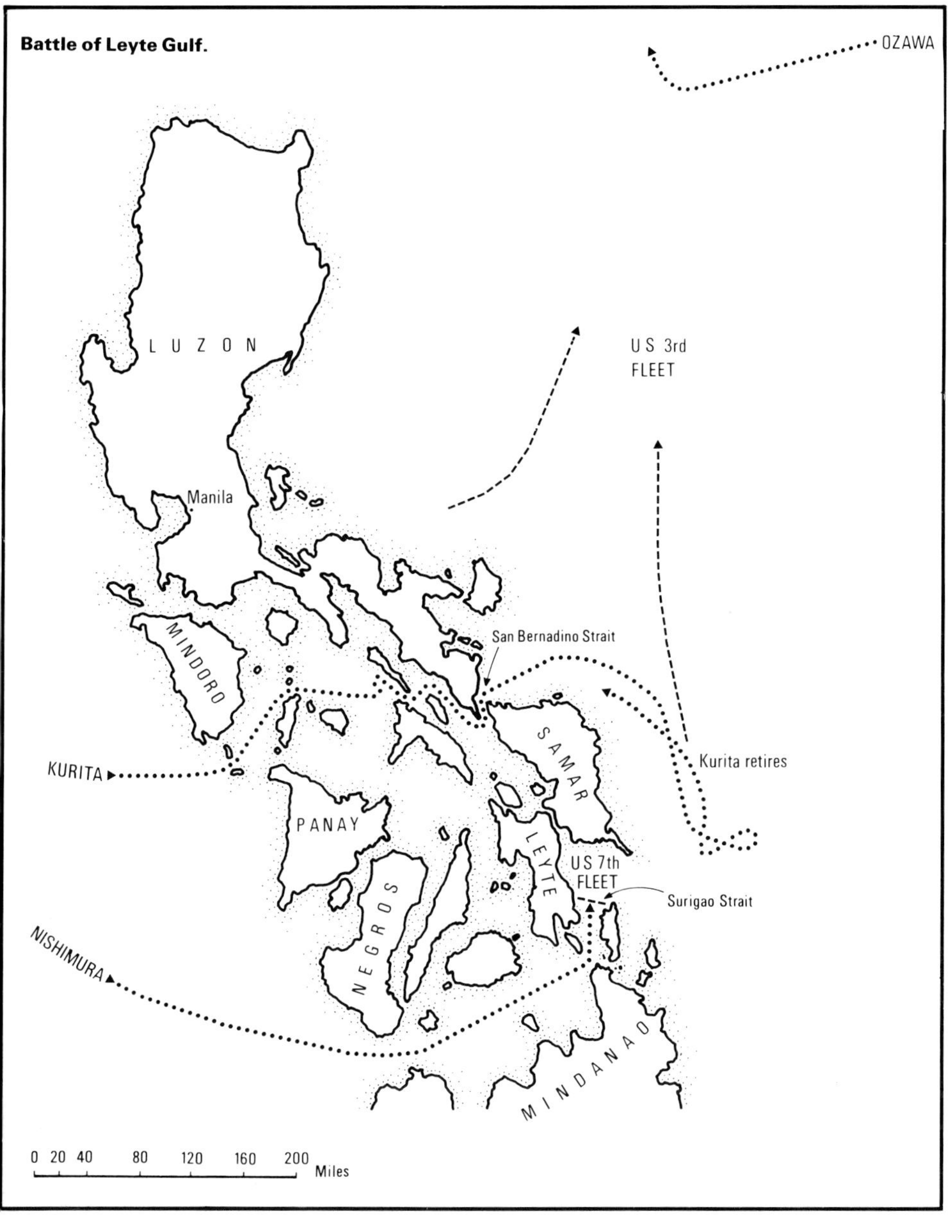

force to be attacked was that under Adm Kurita which included the battleships *Yamato*, *Musahi*, *Nagato*, *Kongo* and *Haruna*. As this force headed eastward in the Sibuyan Sea, they were attacked by aircraft from American carriers. Throughout the day, wave after wave of aircraft attacked the Japanese ships. *Musahi*, one of the 'Yamato' class, was damaged by torpedoes from the third wave. Thereafter the American aircraft concentrated on *Musahi* and she was eventually sunk after more torpedo hits. Kurita's force turned to the west, as if retiring.

Meanwhile, a force under Adm Ozawa, which included the battleships *Hyuga* and *Ise*, was approaching the Philippines from the northeast. The purpose of this force was to act as a decoy, drawing off the US naval forces defending the invasion shipping. Ozawa's group also contained four carriers and was seen as a major threat by the Americans although, in fact, the Japanese carriers carried few aircraft and only had inexperienced aircrews.

The American battleships were divided between two fleets. Adm Halsey's 3rd Fleet contained eight carriers, eight light carriers and the battleships *New Jersey*, *Iowa*, *Massachusetts*, *South Dakota*, *Washington* and *Alabama*, the six most modern battleships in the US Navy. The 3rd Fleet's role was the protection of the invasion forces from attack by Japanese ships or carrier-borne aircraft. In direct support of the invasion, including bombardment, were the battleships of the 7th Fleet, *Mississippi*,

Below:
The tall, pagoda-like structure was characteristic of the older Japanese battleships. The 'Fuso' class of two ships (*Fuso* is seen here, *Yamashiro* was the other) were slow, only being capable of 24kt. They mounted 12×14in guns however, in six turrets, two of them being amidships, one forward, one aft of the funnel. *IWM MH5926*

Maryland, *West Virginia*, *Tennessee*, *California* and *Pennsylvania*, under Adm Kinkaid.

The 3rd Fleet was indeed decoyed by Ozawa's force as the Japanese intended, and the American carriers and battleships headed north to attack the Japanese. Meanwhile, Kurita's force had turned again to the east towards the San Bernardino Straits, where there were no significant US forces to oppose them — the route to Leyte Gulf was open. In addition, a force under Adm Nishimura was heading for the Surigao Straits, a force that included the battleships *Fuso* and *Yamashiro*. However, Nishimura would find his way barred by the US 7th Fleet in the Surigao Straits.

The American forces guarding the Surigao Straits formed three lines of defence. Furthest south were the PT (torpedo) boats, then the destroyers and finally a line of cruisers and battleships. The Japanese force entered the straits after dark, heading north in column: four destroyers, the two battleships and the heavy cruiser *Mogami*. The first attack by American PT boats failed to achieve significant damage. However, the subsequent destroyer attacks shortly after 03.00 achieved considerable success. In particular, the battleship *Fuso* was hit with torpedoes, dropped behind the rest of the force, and sank unseen sometime during the night. The remainder of the Japanese force continued north.

Of the six American battleships positioned across the northern end of the Surigao Straits, *Maryland* and *West Virginia* had 8×16in main armament and the remainder had 12×14in. All the ships were short of armour-piercing shells, having been ammunitioned as a shore bombardment group. However, they had the advantage of fire control radar, *West Virginia*, *Tennessee* and *California* having the most modern. At 03.51 the Japanese ships came within range. The targets were unseen, the night being dark with no moon, but radar provided all the information the Americans needed. The battleship line opened a devastating fire against the Japanese, who stood no chance against the overwhelming American force. It was all over very quickly. At 04.08 on 25 October 1944, *Mississippi* fired the last ever broadside by one battleship against another, 12×14in guns at a range of 19,970yd. *Yamashiro* was sunk, and the supporting Japanese cruisers and destroyers decimated — some sunk — most of the others damaged and retiring as best they could.

Meanwhile, Kurita's force, which still included four battleships including *Yamato*, had exited the San Bernardino Straits and headed south. The only US forces which lay between them and Leyte Gulf, where the invasion shipping could have been at their mercy, was a group of six small carriers escorted only by destroyers. The Japanese, however, did not know how weak this force was and did not know the whereabouts of the main US force. Seen from the Japanese viewpoint, the lack of a clear picture of what was going on was significant:

'Just as day broke at 06.40 on the 25th, and we were changing from night search disposition to anti-aircraft alert cruising disposition (ring formation), enemy carriers were sighted on the horizon. Several masts came in sight at about 30 kilometres to the southeast, and presently we could see planes being launched.

'This was indeed a miracle. Think of a surface force coming up on an enemy carrier group! *Yamato* increased speed instantly and opened fire at a range of 31 kilometres. The enemy was estimated to be four or five fast carriers guarded by one or two battleships and at least 10 heavy cruisers. Nothing is more vulnerable than an aircraft carrier in a surface engagement, so the enemy lost no time in retiring.

'In a pursuit, the only essential is to close the gap as rapidly as possible and concentrate fire upon the enemy. Admiral Kurita did not therefore adopt the usual deployment procedures but instantly ordered "General Attack"; destroyer squadrons were ordered to follow the main body. The enemy withdrew, first to the east, next to the south, and then to the southwest, on an arc-like track. In retreat he darted into the cover of local squalls and destroyer smoke screens, while attacking us continuously with destroyer torpedoes and attack planes.

'Our fast cruisers, in the van, were followed by the battleships, and little heed was paid to co-ordination. Because of the enemy's efficient use of squalls and smoke screens for cover, his ships were visible to us in *Yamato* only at short intervals. The enemy destroyers were multi-funnelled, with high freeboard. Their appearance and torpedo firing method convinced us they were cruisers. We pursued at top speed for over two hours but could not close the gap, in fact it actually appeared to be lengthening. We estimated that the enemy's speed was nearly 30 knots, that his carriers were of the regular large type, that pursuit would be an endless see-saw, and that we would be

Above right:
***Mississippi*, the last battleship to fire at another battleship when she fired her final broadside of 12×14in at *Yamashiro* at 04.08 on 25 October 1944.** *Real Photos*

Right
This ship is of the Japanese 'Yamato' class, almost certainly *Yamato* herself, under attack in the Sulu Sea before she broke through the San Bernardino Straits. These monster ships mounted 9×18.1in guns. *IWM MYF47538*

Far right:
Battleships of the US 7th Fleet of Luzon. These ships sank *Yamashiro* at night in the Battle of the Surgao Strait. These were the older American battleships; this photograph shows *Pennsylvania* leading with a 'Maryland' class astern. *IWM NYF52260*

Above:
***Yamato* could have decimated US forces off Leyte Gulf; however, it was only after the war that the Japanese knew that the way would have been clear to them. On the day, the way appeared to be blocked by a force of fast carriers and battleships. Such is the fog of war.** *Real Photos*

unable to strike a decisive blow. And running at top speed, we were consuming fuel at an alarming rate. Adm Kurita accordingly suspended the pursuit at 09.10 and ordered all units to close. After the war I was astonished to learn that our quarry had been only six escort carriers, three destroyers and four destroyer escorts, and that the maximum speed of these carriers was only 18 knots.' *Tomiji Koyanagi*

Whilst all this was going on, the battleships of the 3rd Fleet were well to the north, about to come within range of Ozawa's force. However, on hearing that Japanese battleships were through the San Bernardino Straits, Adm Halsey detached his six battleships, who immediately headed south at high speed. However, Kurita's force had slipped back through the San Bernardino Straits long before the American battleships could catch them.

Ozawa's force was subsequently attacked by Halsey's carrier-born aircraft and all four of the Japanese aircraft carriers were sunk. The two battleships with the force, *Hyuga* and *Ise*, survived and withdrew with the other remnants of the force. The Japanese had lost three battleships in the Battle of Leyte Gulf, the Americans none. Japanese battleships would mount only one more operational sortie before the war's end. The American ships would continue in support of amphibious operations as Allied forces closed in on the mainland of Japan.

Below:
***Iowa* was detached with the five other battleships of Adm Halsey's 3rd Fleet to attempt to intercept *Yamato*, but the Japanese slipped away through the San Bernardino Straits.** *Real Photos*

8 Indian Ocean II

Meanwhile, the Eastern Fleet had been operating in support of land operations in Burma:

'Probably at the beginning of 1945 [actually in January], two amphibious operations were carried out at Rambre Kyun (we knew it as Ramree Island) and Akyab, both islands off the coast of Burma. These operations were very successful and the bombardment of the capital ships overwhelming. There was very little resistance to the landing troops.

'I was in the port after HA control tower during this action, which started at dawn. The aircraft from the carriers commenced the action, then the battleships opened fire. It is quite an experience to be in an exposed position above X and Y turrets when they open up with full charge. The news over the loud-hailers is coming in reporting the progress of the carrier aircraft, then the noise of gun crews going through their loading drill and the sound of loading machinery. At this point, a pause — ding . . . ding, the firing gongs from the main armament director tower to all turrets — at this point is almost a hush.

'Then comes an absolutely pulverising crashing roar, great masses of orange flame and pungent smoke, all within a great violent shudder as eight guns of four turrets open fire.

'If it is your first time in the open, one feels shaken, senseless, concussed for, I suppose, split seconds. By this time there's the gun crews shouting out again, and in a few more seconds the Ding! Ding! Firing bells are ringing again as the intense noise and firing continues and you realise that the frequency is as about 3 rounds per gun per minute. At about this time you see and hear the guns of other ships, as they follow *Queen Elizabeth* into the bay.' *W. Burley*

Below decks, even away from the noise of the guns, life was not easy:

'In the steamy heat of a control position, 3 or 4 decks below the quarterdeck, clothing quickly became saturated with perspiration. Anti-flash gear worn during Action Stations only added to the discomfort. On several occasions at sea, the fan which supplied air to our port after HACP broke down, and so that the crew would not collapse with heat stroke, the Captain ordered that hatches immediately above the position should be opened until repairs were completed.' *A. Daniels*

'One incident which stands out in my memory with the Eastern Fleet was one day when the whole fleet was manoeuvring at sea, Adm Somerville signalled that, in order to give Warspite's *OOWs experience in station keeping, etc,* Ramillies *was to become the Fleet Guide ship until further notice. This signal was executed, but later in the day the Admiral made a further signal which said that the gyro compass of* Ramillies *had "gone off the board" and was not working correctly. In this connection, the* Ramillies *OOW had not maintained the correct checking procedure with the magnetic compass and, as a result, the fleet had been steering an incorrect course for some time. the flagship then made a signal to* Ramillies; *"Indicate name of officer concerned", and from both main yardarms was hoisted the name of the unfortunate officer concerned for the whole fleet to see.'*

Below:
***Warspite* leading units of the Eastern Fleet.**
IWM A11788

9 British Pacific Fleet

Queen Elizabeth was to remain the mainstay of the Eastern Fleet in the Indian Ocean. On 4 February 1945 *King George V* arrived in Fremantle, Australia, to join the British Pacific Fleet, the nucleus of which was four aircraft carriers. From now on, *KGV* would operate in the Pacific in co-ordination with American forces, to be joined by *Howe* in the spring of 1945. Australia formed the rear base for the fleet but successful operations were dependent on replenishment at sea:

'Ammunitioning, storing and refuelling were normally carried out in harbour, but whilst in the Pacific (on more than one occasion) we were at sea for periods of about three months at a time, which necessitated all these operations being undertaken many times at sea. Evaluations had taken place in UK waters for refuelling at sea, but I am sure that a lot was learned as we went along in the Far East. The supply organisation was most efficient. We rarely went short of fresh vegetables, etc, and UK mail was received regularly and arrived within days rather than weeks, even in the remote areas of the Pacific.' *J. E. Bowen*

Prior to commencing operations with the Americans, the British Pacific Fleet gathered in Australia. Before going on to look briefly at the final operation of the war, let us pause for a moment to look at some more aspects of the life of the officers in a battleship, starting in the wardroom:

'The wardroom was a palatial place and beautifully equipped and furnished, so

Above:
A tranquil scene — *Howe* on her way through the Suez Canal to join the British Pacific Fleet. *Howe* and *King George V* were the nucleus of the British Pacific Fleet, along with the carriers. *G. A. Langridge*

Right:
***Howe* on her way to the Far East.**
G. A. Langridge

tastefully too. We all took great pride in our wardroom and saw to it that it was kept in good order. We also had a very good ante-room on the port side, as one came in from the quarterdeck. In the Indian Ocean, in the Pacific and in waters around Australia, whilst at sea we wore khaki shorts and sandals plus a khaki cap cover. We hung our shirts up on our pegs in the ante-room lobby and put them on before entering the ante-room and wardroom. We never abused this privilege given to us by the Captain who, I remember, ran *Anson* more like a destroyer when at sea. We had Wardroom Attendants who looked after their own particular officers with regard to laying out uniforms, what to, where and when. These were Royal Marines — full of guile and cunning! They were literally our saviours to us RNRs and RNVRs, who taught us ship's protocol in a battleship. They took a pride in turning out their young charges. We, in turn, learned a great deal from them.'
J. V. Haddock

Life for the Captain, paradoxically, was not so comfortable, certainly at sea:

'The Captain took over the Admiral's quarters [when there was no Admiral

Above:
The view from *Howe* as she refuels by the astern method from the tanker ahead, whilst a cruiser refuels by the abeam method.
G. A. Langridge

Left:
***King George V* arriving in Australia where the British Pacific Fleet was built up and where the main base for the fleet was established.**
J. E. Bowen

Below:
***Arkansas* (BB-33), the oldest US battleship to serve in World War 2. She spent much of her time in the Atlantic but was present in the Pacific for the attacks on Iwo Jima and Okinawa.** *Real Photos*

Table 1

Date	Duty	Route or AT	Mileage	Days at Sea	Average Speed	Fuel Burnt	Ammunition Expended 14in	5.25in	Pp	Bof	Oerl
31 Jul *27 Oct*	Working Up	Liverpool-Scapa	4,693.7	9	–	5,491	162	617	4,482	763	9,900
28 Oct *11 Nov*	Passage –	Scapa-Alex	3,956.1	9	18.3	3,537	–	96	347	94	928
11 Nov *13 Nov*	In Harbour	–	–	–	–	–	–	–	–	–	–
13 Nov *15 Nov*	OPs Milis	Alex	927.8	2	18.6	} 1,725	59	279	726	147	419
15 Nov *30 Nov*	In Harbour	Alex & Exercises	1,026.5	4	14.9		–	–	–	–	–
01 Dec *15 Dec*	Passage	Alex-Trinco	4,409.3	14	18.8	4,161	–	–	1,115	–	–
15 Dec *15 Jan*	In Harbour	Trinco & Exercises	718.9	2	16.0	743	–	144	1,496	188	2,180
16 Jan *04 Feb*	Operation Meridian	Trinco-Fremantle	8,309	19	18.5	5,905	–	144	629	24	1,137
04 Feb *05 Feb*	In Harbour	–	–	–	–	–	–	–	–	–	–
11 Feb *11 Feb*	Passage	Fremantle-Sydney	2,515	6	17.9	2,055	11	25	28	–	–
11 Feb *28 Feb*	In Harbour	Sydney	–	–	–	–	–	–	–	–	–
28 Feb *07 Mar*	Passage	Sydney-Manus	2,901	8	16.2	2,450	–	54	287	42	–
07 Mar *18 Mar*	In Harbour	Manus & Exercises	133	2	16.6	191	–	–	–	–	–
18 Mar *20 Mar*	Passage	Manus Ulithi	865.9	2	16.6	794	–	105	–	–	–
20 Mar *23 Mar*	–	Ulithi	–	–	–	–	–	–	–	–	–
23 Mar *23 Apl*	OP Iceberg	Ulithi Leyte	12,147	31	16.9	10,387	–	119	1,930	118	2,183
23 Apl *01 May*		Leyte	–	–	–	–	–	–	–	–	–
01 May *28 May*	OP Iceberg	Leyte-Guam	10,874	27	16.1	8,748	77	188	2,190	212	2,013
28 May *30 May*		Guam	–	–	–	–	–	–	–	–	–
30 May *06 Jun*	Passage	Guam-Sydney	3,150	6	20.6	3,291	–	94	1,641	193	391
06 Jun *28 Jun*		Sydney & Exercises	313.4	1	13.3	249	–	142	2,784	342	2,635
28 Jun *03 Jul*	Passage	Sydney-Manus	2,231	6	15.5	1,626	–	352	5,083	1,149	1,563
03 Jul *06 Jul*		Manus	–	–	–	–	–	–	–	–	–
06 Jul *27 Aug*	OPs	Manus-Sagami	19,200	52	15.5	15,086	541	634	2,144	360	2,015
27 Aug *31 Aug*		Sagami	–	–	–	–	–	–	–	–	–
31 Aug	Passage	Sagami-Tokyo	44	–	11.0	130	–	–	–	–	–
Thence operations in Far East until (various practices):											
01 Mar 1946		Portsmouth									
	Grand Totals		78,415.6	200	16.5	66,568 tons	850	2,993	24,882	3,632	25,364

embarked] and they were massive, but the only time he spent in them was in harbour. At sea he lived in a small shelter about three steps down from the compass platform, and it was small — just a bunk and a small table. The PO Steward and myself were messed in the Captain's Mess and lived in the pantry. At sea I used to get my head down in the spare cabin.

'At sea I must have been up and down the bridge thousands of times, at least six times a day with meals and other requirements, and spent many hours up there waiting for the Captain whenever there was a flap on, and there were many.'
D. K. Bean, Captain's steward

At sea the ship was run from the bridge but, in harbour, the Officer of the Watch was on the quarterdeck:
'OOW on the quarterdeck was always extremely hectic, as boats were always coming and going and pipes and bugle calls had to be sounded at various times of the day. When we finished our stints on the quarterdeck of *Anson* as OOW, we felt shredded, with so many things to see and to remember. *J. V. Haddock*

The Australians in those days were so keen to have anything as a souvenir. The sailors and ourselves had to guard all our personal items. I well remember when, as Officer of the Watch in harbour at Woolamaloo, Sydney. having inspected libertymen for shore leave, some 10 minutes later one rating came back on board, literally wearing only his underpants! The 'Welcome Australia' folk had cleaned him of all his uniform as keepsakes.

The Admiral must have been a very lonely man. Most evenings he would appear on the quarterdeck for his evening constitutional, walking slowly up and down, his head bowed in thought. Occasionally when I was Bosun's Mate, he would ask the OOW if he could spare me for a while to join him. We would talk of life at home — mine and his — my aspirations, all those little things that men talk about and without the vast difference in rank interfering. 'Leave out the sirs', he told me. 'I just want to chat.' I would be dismissed with a quiet 'Thank you, Bosun's Mate, back to your duties', and his slow pacing would continue.

The stay in Australia was not for long, and thereafter the ships in the Pacific Fleet were to spend long periods at sea.

Table 1 shows *King George V's* programme between 18 March and her arrival in Tokyo at the end of August 1945. During this period, she supported operations off Formosa and bombarded the mainland of Japan near Tokyo on two occasions. The main threat which battleships faced during the final assaults on Iwo Jima and Okinawa on the way to the Japanese mainland were the Kamikaze suicide bombers:
'The suicide bombers certainly attacked us. There was one particular day that I well remember when the *Howe* and *KGV* were called over to attack an island with some particular airfields and this we did, and we left the carriers and the cruisers. We took some destroyers with us. We did the job and, as we sailed back in the late afternoon, it was an incredible sight that met our eyes. The carriers . . . some of them were alight . . . wreckage was strewn all over the flight decks. They had been attacked by suicide bombers . . . It wasn't over by any means. They went hell for leather for us and we were absolutely delighted to be able to shoot one down, which was actually aimed at the bridge, but which went between the foremast and the mainmast, below the radio aerials, and came down off the starboard quarter.' *D. Higgins*

Left:
Table 1 HMS *King George V's* programme between 31 July 1944 and her arrival in Tokyo at the end of August 1945.

Below:
The main purpose of battleship operations in the latter stages of the war was to provide support for amphibious landings. Here an American battleship lies off the beaches as waves of landing craft shuttle to and fro.
IWM OWIL 217702

Dates of particular interest

28 Oct '44	Left Scapa
04 Feb '45	Arrived in Australia
23 Jan '46	Left Australia
15 Aug '45	Cessation of Hostilities
27 Aug '45	Anchored in Sagami Bay
31 Aug '45	Anchored in Tokyo Bay
02 Sep '45	Official Surrender of Japan
01 Mar '46	Arrived in Portsmouth

Bombardments

14 Nov '44	Milos
04 May '45	Miyako
04 May '45	Hiraari Airfield
17 Jul '45	Hitachi
29 Jul '45	Hammamatsu

Air Strikes

Operation 'Meridian'	
24 Jan '45	Against Palangang
29 Jan '45	(*KGV* shot down 1 aircraft)
Operation 'Iceberg'	
26, 27, 31 Mar '45	Sakashima
1, 6, 7, 12, 13, 16, 17, 20 April '45	Gunto
4, 5, 8, 9, 12, 13, 16, 17, 20,	Formosa
21, 24, 25 May '45	Formosa
Operations with US 3rd Fleet	
17, 18, 24, 25, 27, 29, 30 July '45	
8, 9, 10, 13, 15 Aug '45	
Emergencies	
Flash Reds	49
Flash Blues	86

Table 1 also shows the enormous amounts of fuel used and also the ammunition expenditure. This all had to be replenished at sea:
'We [*Howe*] spent 3 days shelling, then came out to refuel at sea, meeting a "Fleet Train". We transferred stores and personnel by breeches buoy, took on mail, etc, then back again for another three days. This was in support of the American Navy at Okinawa.'

King George V had the honour of being in Tokyo Bay for the formal surrender of Japan on 2 September 1945. She had also been the last British battleship to fire her guns in anger when she had bombarded Hammamatsu, south of Tokyo, on 29 July. But there was still work to be done as *Anson*, a late arrival, found:
'. . . Then on our way from Singapore, Fremantle and Sydney, arriving there for VJ-Day! At Sydney there were great celebrations and I was in the victory march. Next we provisioned ship and were on our way to relieve Hong Kong and release POWs at Stanley Camp. The Japs were still on Victoria Peak when we arrived and the Royal Marines aboard us were put ashore to restore the peace. I was with a party put aboard a Jap destroyer to salvage her as she had been sabotaged. This, I'm pleased to say, we achieved after a struggle.'
E. D. Sharpe

It was all over apart from the clearing up. The last battleships from the Far East arrived back in the UK in July 1946.

Above:
HMS *King George V's* activities 1944-45.
Below:
***Anson* lying in Victoria Harbour, Hong Kong, where she dealt with the final pockets of Japanese resistance and started the process of liberating the prisoners of war.** *IWM A30746*

Nine

Post-1945

1 Korea

After World War 2 the battleship fleets of the world were very quickly reduced. In the Royal Navy, only *Vanguard* remained operational after 1950, all the others having gone to scrap. The French retained *Richelieu* and *Jean Bart* and they continued in service into the 1950s. The German battleships were no more. Only *Nagato* survived the war in the Japanese Navy but was expended in atomic bomb tests. A number of US battleships were also expended in such tests, notably at Bikini Atoll; many others, after periods in reserve, went to scrap, although some have been preserved as museum ships.

The only ships which have been in action since 1945 are the French *Jean Bart* and the four ships of the 'Iowa' class in the US Navy. Of the latter *Iowa*, *New Jersey* and *Wisconsin* went into reserve after World War 2 and were 'mothballed'. In this process much of the ship and her equipment is sealed up in air-tight covers, with the atmosphere within kept dry. In this way equipment can be preserved for long periods of time and a ship can be re-activated, if required for service, without large numbers of men having to be employed preserving and maintaining the ship in the meantime. *Missouri* was kept in service as a training ship.

The Korean War saw the re-activation of the 'Iowa' class. One ship at a time was deployed to Korean waters to act as a task force flagship, and to provide fire support for ground forces ashore. Shore bombardment, or 'naval gunfire support' as it became known, had been the primary role of the battleships in the latter part of World War 2. The advantage of using a ship in this role is that it can stay 'on task' much longer than a ground-support aircraft, and can often reach further behind enemy lines than shore artillery by being able to operate from the seaward flank of the enemy land forces. In addition, a battleship in particular has much greater fire power than an artillery battery, 155mm or 6in being the normal maximum

ight:
***owa* in action off the coast of Korea. The four hips of the 'Iowa' class were all re-activated or operations off Korea.** *US Navy*

size of gun available to shore artillery against the 16in of the 'Iowa' class.

Between them, the four ships of the 'Iowa' class provided one ship on station off Korea between September 1950 and March 1953, each ship doing one deployment of about six months, except for *Missouri* who did two deployments. The class went into reserve at various intervals after the end of the Korean War and all four ships were in reserve in 1958.

Meanwhile, the French *Jean Bart* was present off Suez in 1956 during the joint Anglo-French landings. However, as the last of the French battleships, she was decommissioned in 1961. Thus by 1961 there were no battleships in commission in any navy.

Right:
Shells actually leaving the barrels of the 16in guns on *Iowa* (BB-61) during a fire-power demonstration. *US Navy*

2 Vietnam

However, in 1968 *New Jersey* was recommissioned for operations off Vietnam. There was a shortage of big-gun cruisers to provide naval gunfire support, so the battleship was re-activated:

'The ship left Philadelphia in the early part of '68 . . . and went from Philadelphia down to Norfolk, Virginia, and they had ceremonies as the ship was leaving . . . At Norfolk we were the world's only active battleship and, of course, Norfolk is the Headquarters of the US Atlantic Fleet. When we were walking around the town with our patches on saying "USS *New Jersey*", we were really special people. And the Captain of the ship, Capt J. Edward Schneider, led us all to believe that we were really the cream of the crop that the Navy had to offer . . . All of the guys that were working on the ship were really bright guys and it didn't matter what their jobs were — yeomen or personnelmen, supplymen, radarmen, electricians, gunners mates, boatswains — these guys were probably the best the Navy had to offer. All were congregated on this one ship which really made life pretty good. . . . It was really an experience being on the ship.' *P. Siegart*

After working-up *New Jersey* crossed the Pacific to Subic Bay in the Philippines, where stocks of 16in ammunition were held. From there she sailed for Vietnam in September 1968:

'We left Subic Bay to go to Vietnam. I'll never forget the day before we were arriving off Vietnam, and it came over the loudspeaker that tomorrow we would be in Vietnam. None of us knew what to expect and I went to sleep that night and I borrowed a helmet from some guy. We didn't have helmets in the Navy, but I remember switching bunks with someone else as I wanted to be behind one of the chutes that hoisted shells up to the 5in guns. That thing was so thick. I was afraid that if any machine gunfire or shells came through the side of the ship, at least I would have this huge chute in front of me to block any of the bullets. So I went and I got a flak jacket and I slept with a flak jacket on that night.

'I didn't get much sleep, but at 6 o'clock in the morning when they announced our location and the temperature, they said we were off the coast of Vietnam. I had on more stuff than that to stop bullets, and I remember that the porthole was closed so I went to open it up very slightly but I couldn't see anything. So I decided to open up the door and go out on deck, looking like someone with an armoured suit. I went out on the deck and it was like being off the coast of Florida. There were people in boats waving, there were Navy personnel in little speed boats with swimming trunks on, and I said "I don't believe it!". So I looked like a damned fool and I took off a whole lot of stuff. The essence of it was that, in those days, if you were south of the De-Militarised Zone on the water, except maybe in the river patrol boats, then you were very, very safe.' *P. Siegart*

However, it was only a few days before *New Jersey* went into action:

'We travelled north of the DMZ and we started operations off the northern part of Vietnam. The first action we really saw . . . there was a bunch of little barges on

Left:
Replenishing ammunition on board *New Jersey* during 1986. *US Navy*

something, maybe six or seven of them, travelling along the coast of Vietnam. We were about a mile out when these things were spotted. They obviously had no chance against a battleship and we expended so much ammunition blowing these things up, whatever they were, that we got a message from 7th Fleet telling us not to waste ammunition. The key really is that the North Vietnamese had nothing on the water that could do anything to us. There were MiGs in the air but the MiGs never really approached the ship — they were always inland because the aircraft carriers were out in the Tonkin Gulf and there were always American fighters very close by; we always knew that if any MiGs came the fighters would go after them. The MiGs never really did anything although we would sit off the coast and shoot shells into North Vietnam. I remember that we were fired at — I think the shells were coming from a Chinese island in the Tonkin Gulf. A shell came and exploded maybe half a mile from the ship and, because the *New Jersey* could shoot well over 20 miles, the Captain immediately ordered the ship to go out beyond the range of whatever was shooting at us, and then started shooting back.' *P. Siegart*

New Jersey spent 120 days in operations off South Vietnam, the longest period at sea being some 47 days. Over 5,000 rounds of 16in were fired and about 15,000 rounds of 5in. In the gun magazines, things did not always go well:

'Personnelmen and yeomen [men employed on secretarial and administration duties] are people that are generally sent down into the magazines where all the shells are. It was our job to get the particular kind of shell and send it up the hoist to the guns. There was one time when we were woken up around midnight because it was our tour of duty; we would do 4 hours on duty. I was really tired and I was sent down there. They usually tried to have one gunner's mate with you who knows what's going on. We were inside and someone put a shell in the hoist. I don't think the shell was put in correctly because all of a sudden there was a banging against the shell. 3 personnelmen, 3 yeomen and 1 gunner's mate were down there and when that thing started clanging — some mechanical part was hitting the shell — needless to say the personnel men and the yeomen were out of that magazine so fast! But the gunner's mate, who had experienced this and knew what was going on, him and maybe a boatswain's mate who he yelled to, fixed the machine. We were really not trained outside of lifting up those shells and sticking them in the hoist. We didn't know much about combat. Since were were not trained on how those machines operated, it was not surprising that we took off.' *P. Siegart*

New Jersey was withdrawn from operations off Vietnam in March 1969 as part of the general reduction in US naval activity in the area. The war zone itself at sea was, in practice, a very small area and the contrast between operations off Vietnam and visits to nearby ports was extreme. After days spent shelling positions in North Vietnam, a few days steaming away, the crew could be doing their shopping:

'When we went to Japan for R&R (Rest and Relaxation), everyone had a lot of money because everyone was paid in cash at sea. Everyone bought all kinds of electronic equipment, and spent a lot of time playing with that kind of stuff on board.' *P. Siegart*

By the end of 1969, *New Jersey* had been decommissioned and all four ships of the 'Iowa' class were, once again, back in reserve.

Below:
A new dimension to battleship operations — operating helicopters, in this case a UH-1 Iroquois. The photograph gives a very good impression of the sheer size of the quarterdeck on an 'Iowa' class battleship. *US Navy*

3 Re-activation of the 'Iowa' class

The 'Iowa' class remained in reserve for a further 13 years. However, in the early 1980s it was decided to re-activate the class once again. For the Korean and Vietnamese wars few modifications were incorporated in the ships; the re-activation of the class in the 1980s included significant modifications. The roles of ships were to include acting as the predominant unit of a surface action group, either as part of a carrier battle group or independently, or to provide naval gunfire support to amphibious operations or any US forces ashore. The main armament was retained but, in addition, a strong missile armament has been fitted. 4×Quad mountings for Tomahawk cruise missiles have been fitted amidships with a further 4×Quad on the after superstructure. Also Quad launchers have been fitted either side of the after funnel for Harpoon anti-shipping missiles.

New Jersey recommissioned in December 1982, with *Iowa* following in 1984. *Missouri* became operational in 1986 and *Wisconsin* is due to follow in 1987/88. Within a year of recommissioning, *New Jersey* was back in action, this time off the coast of Lebanon.

On 25 September 1983 *New Jersey* joined the naval units providing support for the Multinational Peacekeeping Force in Lebanon, with operations centred on Beirut. On 23 October a truck carrying explosives was detonated inside the US Marine compound in Beirut, killing 241 US personnel. *New Jersey* provided medical support and landed her own Marine detachment to assist ashore. Gunfire support for forces ashore was requested for the first occasion since the Vietnam war on 14 December, when *New Jersey* fired 11×16in at anti-aircraft sites ashore which had been engaging US reconnaissance aircraft. The next day the 5in engaged Druze militia positions around Beirut. The ship remained on station over Christmas but visited Haifa for New Year.

New Jersey returned to Lebanon on 3 January and continued to provide gunfire support for the forces ashore. The longest shoot occurred on 8 February when, over a period of almost half a day, 288 16in rounds were fired into positions in the Chouf Mountains behind Beirut. *New Jersey* was withdrawn from operations off Lebanon on 2 April 1984.

Since recommissioning, *Iowa* and *New Jersey* have taken part in many exercises.

Below:
New Jersey after recommissioning in 1982. The Tomahawk or cruise missile launchers are the flat boxes just above the two twin 5in turrets; the Harpoon launchers are immediately above the Tomahawk launchers and they are in two groups of four set at an angle. *US Navy*

Right:
A necessary skill for today's battleship — replenishing 16in ammunition at sea from the supply ship *Wabash*. *US Navy*

Below right:
***New Jersey*, seen here firing a 16in salvo, was back in action off Lebanon in 1983/84 supporting US Marines ashore who formed part of the Multinational Peacekeeping Force.** *US Navy*

Iowa is based on the east coast of the USA while *New Jersey*, despite her deployment to Lebanon, is in fact based in Long Beach, California. *Iowa* has been seen in major NATO exercises, notably 'Ocean Safari 85' when she acted as the prime unit of the Battleship Battle Group, co-ordinating surface warfare activity for carrier forces.

The armoured ship still has a role to play. With aircraft today forced to use stand-off weapons because of the anti-aircraft missile defences fitted in ships, the ships of the 'Iowa' class are well able to withstand the damage which missiles like Exocet can cause. Their own missile armament, Tomahawk and Harpoon, is impressive; their guns still have a role to play in certain scenarios, supporting land operations. However, the battleship became vulnerable to the submarine the day the first operational submarine went to sea in the early 1900s. They will always require strong ASW forces to protect them. Nevertheless, it is likely that the 'Iowa' class ships will be around for some years to come.

Left:
***New Jersey* firing a Tomahawk cruise missile during test firings off the coast of Southern California in 1983.** *US Navy*

Below:
Echoes of the past — two US battleships together at sea. *Missouri* is seen under the guns of *New Jersey*. *US Navy*

Bottom:
The battleship as part of a modern task force. The low lines of the 'Iowa' class are evident when compared with the cruiser on her starboard beam. *US Navy*

Bibliography

Battle of the Falkland Islands

Coronel and the Falklands, J. Irving (A. M. Philpot c1925)
Coronel and Falkland, B. Pitt (Cassell 1960)
Defeat at the Falklands, E. P. Hoyt (Robert Hale 1956)
The Pursuit of Admiral von Spee, R. Hough (G. Allen & Unwin 1969)

Dardanelles

Dardanelles, Denham
Dardanelles Dilemma, E. K. Chatteron (Rich & Cowan 1935)
The Sailor's War, P. H. Liddle (Blandford 1985)

Battle of the River Plate

The Battle of the River Plate, D. Pope (W. Kimber 1956)
Graf Spee, W. Powell (Hodder & Stoughton 1955)

Battle of North Cape

The Battle of North Cape, Lt-Cdr M. Ogden (W. Kimber 1962)
The Drama of the Scharnhorst, Fritz-Otto Busch (Robert Hale 1956)
The Death of the Scharnhorst, J. Winton (Anthony Bird 1983)

German Navy

The German Navy in World War II, E. von de Posten (A. Barker 1970)
The German Navy in World War II, Jak P. Mallmann Showell (Arms & Armour Press 1979)
German Capital Ships, P. Beaver (Patrick Stephens 1980)

US Ships

Battleships of the US Navy in World War II, Terzibaschitsch (Brasseys 1977)

Japanese Navy

The End of the Japanese Navy, Masanori Ito (Weidenfeld & Nicholson 1956)
The Japanese Navy in World War II, various (US Naval Institute 1969)
Warships of the Imperial Japanese Navy, Jentschura, Jung & Mickel (Arms & Armour Press 1977)

Individual Ships

Battleship Barham, G. Jones (W. Kimber 1979)
Battleship Bismarck, Von Mullenheim-Rechbug (Bodley Head 1981)
The Mighty Hood, E. Bradford (Hodder 1959)
Scharnhorst & Gneisenau, R. Garrett (David & Charles 1978)
The Battlecruiser Hood, J. Roberts (Conway Maritime Press 1982)

Technical Details of Ships

Battleships, A. J. Watts (Macdonald & Janes 1978)
Fighting Ships of World War I *and World War II*, compilers A. Maclean & S. Poole (Phoebus 1976)
Battleships and Battlecruisers, B. R. Coward (Ian Allan 1986)
British Battleships, O. Parkes (Seeley Service 1957)
British Battleships of World War II, Raven & Roberts (Arms & Armour Press 1976)
An Illustrated Guide to Battleships & Battlecruisers, J. Jordan (Salamander 1981)

General

Sunk, D. Woodward (G. Allen & Unwin 1982)
Great Battle Fleets, O. Warner (Hamlyn 1973)
Guns at Sea, P. Padfield (Hugh Evelyn 1973
The Battleship Era, P. Padfield (R. Hart Davis 1972)
The Narrative of a Naval Nobody, D. Fairburn (J. Murray 1929)
Convoy, J. Winton (M. Joseph 1983)
A Sailor's Odyssey, Admiral of the Fleet Viscount Cunningham (Hutchinson 1951
The Navy at War, Capt S. W. Roskill (Collins 1960)

World War 1

The personal accounts were all drawn from material held in the 1914-18 Archives at Sunderland Polytechnic, collected and administered by Peter H. Liddle FR Hist Soc. Accounts from the following have been used:
Arkayan, Askir, General: Turkish battery officer; Banks, H. Cdr: Midshipman, *Agamemnon*; Berridge, W. L.: *Albion*; Bramham: *Canopus*; Brand, E. S. Capt: Sub-Lieutenant, *Valiant*; Bullock, E. W. Chief Stoker: *Inflexible*; Clarke, A. W. Capt: Midshipman, *Implacable*; Davis, Adm Sir William: Midshipman, *Neptune*; Dick, R. M. Rear-Adm: Midshipman, *Carnarvon*; Sub-Lieutenant, *Valiant*; Drage, C. Cdr: Midshipman, *London*; Fox-Pitt, T. S. Cdr: Midshipman, *Lord Nelson*; Keeler, G. E.: Able Seaman, *Lord Nelson*; Nichols, C. A. G. Capt: Midshipman, *New Zealand*; Ouvry, J. G. D., Cdr: Midshipman, *Tiger*; Seyeaux: Midshipman, *Gaulois*; Thomas, A. W.: Able Seaman *Malaya*.

The author is most grateful to Peter Liddle for access to the above sources.

World War 2 et seq

The following individuals provided materia
G. T. Avery, J. G. Barlow, D. K. Bean, N. H. Bissell, J. E. Bowen, W. E. Broomfield, A. W. R. Brown, W. Burley, B. C. Cambray, A. Daniels, Lt G. C. Davi R. G. Duquemin, C. B. Elsmore, R. Emmington, A. V. Godding, Lt Cdr J. V. Haddock, W. Handley, D. Higgins, N. Hill, G. A. Langridge, E. R. Moyle, I. R Nicholls, S. Pearson, Lt C. T. Piddington, T. W. Pope, R. W. Scott, P. W. Siegart, E. D. Sharp, D. Smith, R. V. Racey, E. C. Taylor, F. Troughton, B. Wainwright, W. C. West, G. Whyte.